Ultra and Extremely Low Frequency Electromagnetic Field

Ultra and Extremely Low Frequency Electromagnetic Field

Editor

Namrata Saini

Ultra and Extremely Low Frequency Electromagnetic Field

Edited by **Namrata Saini**

Printed in 2017

ISBN: 978-1-68117-159-3

Library of Congress Control Number: 2015936572

© 2016 by
SCITUS Academics LLC,
616, Corporate Way, Suite 2, 4766,
Valley Cottage, NY 10989

www.scitusacademics.com

Contents

Preface vii

Chapter 1   Application of Polarization Ellipse Technique for Analysis of
             ULF Magnetic Fields from Two Distant Stations in Koyna-warna
             Seismoactive Region, West India 1

             F. Dudkin, Gautam Rawat, B. R. Arora, V. Korepanov, O. Leontyeva,
             and A. K. Sharma

Chapter 2   Phenomena of Electrostatic Perturbations before Strong
             Earthquakes (2005–2010) Observed on DEMETER 23

             X. Zhang, X. Shen, M. Parrot, Z. Zeren, X. Ouyang, J. Liu, J. Qian,
             S. Zhao, and Y. Miao

Chapter 3   Possible Electromagnetic Effects on Abnormal Animal Behavior
             before an Earthquake 43

             Masashi Hayakawa

Chapter 4   A Deterministic Approach to Earthquake Prediction 63

             Vittorio Sgrigna and Livio Conti

Chapter 5   Electromagnetic Radiation Energy Harvesting –The Rectenna
             Based Approach 119

             Gabriel Abadal, Javier Alda, and Jordi Agustí

Chapter 6   Features of Usage of Electromagnetic Field of Extremely Low
             Frequency for the Storage of Agricultural Products 157

             Kasyanov Gennady Ivanovich, Syazin Ivan Evgenyevich, Grachev
             Alexandr Vasilyevich, Davidenko Taisiya Nikolaevna, and
             Vazhenin Evgeniy Igorevich

Chapter 7   Theoretical Modeling and Experimental Analysis of Drying
             Process in Electromagnetic Field 173

             Arif Memmedov, Teymuraz Abbasov, and Mustafa Şeker

vi

Chapter 8   Compound Auroral Micromorphology: Ground-based High-
             speed Imaging 199
             Ryuho Kataoka, Yoko Fukuda, Yoshizumi Miyoshi, Hiroko
             Miyahara, Satoru Itoya, Yusuke Ebihara, Donald Hampton,
             Hanna Dahlgren, Daniel Whiter, and Nickolay Ivchenko

Chapter 9   Study of Pc1 Pearl Structures Observed at Multi-Point Ground
             Stations in Russia, Japan, and Canada 217
             Chae-Woo Jun, Kazuo Shiokawa, Martin Connors, Ian Schofield,
             Igor Poddelsky, and Boris Shevtsov

             Citations..243
             Index..247

Preface

Extremely low frequency (ELF) is the ITU designation for electromagnetic radiation (radio waves) with frequencies from 3 to 30 Hz, and corresponding wavelengths from 100,000 to 10,000 kilometers. In atmospheric science, an alternative definition is usually given, from 3 Hz to 3 kHz. In the related magnetosphere science, the lower frequency electromagnetic oscillations (pulsations occurring below ~3 Hz) are considered to lie in the ULF range, which is thus also defined differently from the ITU radio bands. The major emphasis of this book is on physical mechanisms and sources of the ULF/ELF natural electromagnetic fields noises. One of the challenges of this research is to fully understand electromagnetic effects and physical processes in the rocks deep in the Earth's crust.

Editor

Application of Polarization Ellipse Technique for Analysis of ULF Magnetic Fields from Two Distant Stations in Koyna-warna Seismoactive Region, West India

F. Dudkin[1], Gautam Rawat[2], B. R. Arora[2], V. Korepanov[1], O. Leontyeva[1], and A. K. Sharma[3]

[1]Lviv Centre of Institute of Space Research, Lviv, 79000, Ukraine
[2]Wadia Institute of Himalayan Geology, Dehradun, 248 001, India
[3]Department of Physics, Shivaji University, Kolhapur, India

ABSTRACT

A new approach is developed to find the source azimuth of the ultra low frequency (ULF) electromagnetic (EM) signals believed to be emanating from well defined seismic zone. The method is test applied on magnetic data procured from the seismoactive region of Koyna-Warna, known for prolonged reservoir triggered seismicity. Extremely low-noise, high-sensitivity LEMI-30 search coil magnetometers were used to measure simultaneously the vector magnetic field in the frequency range 0.001–32 Hz at two stations, the one located within and another _100 km away from the seismic active zone. During the observation campaign extending from 15 March to 30 June 2006 two earthquakes (EQs) of magnitude (M_L >4) occurred, which are searched for the presence of precursory EM signals.

Comparison of polarization ellipses (PE) parameters formed by the magnetic field components at the measurement stations, in select frequency bands, allows discrimination of seismo-EM signals from the natural background ULF signals of magnetospheric/ionospheric origin. The magnetic field components corresponding to spectral bands dominated by seismo-EM fields define the PE plane which at any instant contains the source of the EM fields. Intersection lines of such defined PE planes for distant observation stations clutter in to the source region. Approximating the magnetic-dipole configuration for the source, the magnetic field components along the intersection lines suggest that azimuth of the EM source align in the NNW-SSE direction. This direction well coincides with the orientation of nodal plane of normal fault plane mechanism for the two largest EQs recorded during the campaign. More significantly the correspondence of this direction with the tectonic controlled trend in local seismicity, it has been surmised that high pressure fluid flow along the fault that facilitate EQs in the region may also be the source mechanism for EM fields by electrokinetic effect.

INTRODUCTION

Short-term earthquake (EQ) prediction, despite intensive efforts in last half a century, still remains unattainable though numbers of promising leads and directions are indicated (see Uyeda et al., 2009, for recent review on the subject). The anomalous electromagnetic (EM) emission in ultra low frequency (ULF) band (0.001–10 Hz), believed to be emanating from within the focal zones, have emerged as potential precursor candidates for short-term EQ prediction (Hayakawa et al., 1996, 2000, 2004, 2007; Molchanov and Hayakawa, 1995; Molchanov et al., 1992, 2004). This observational conviction is further reinforced from the suggestions that mechanical deformations or microfracturing in the impending focal zones may give rise to pre- and/or co-seismic EM emission in ULF band due to one or more of the following factors: (1) inductive effect resulting from the movement of conductive medium in the Earth's permanent magnetic field [Surkov, 1999; Surkov et al., 2003]; (2) displacements of boundaries between high and low conductive crustal blocks (Dudkin et al., 2003); (3) electrokinetic effect (Mizutani et al., 1976; Fitterman, 1979; Fedorov et al., 2001); (4) piezoelectric orpiezomagnetic effects (Martin et al., 1978; Ogawa et al., 1985; Johnston et al., 1994; Ogawa and Utada, 2000) and (5) microfracture electrification (Molchanov and Hayakawa, 1995) (all references are given as example). The underground ULF EM field attenuates only little in crustal material and hence on theoretical consideration associated magnetic field can be detected to a large distances up to 100–150 km (Hayakawa et al., 2007).

The practical detections and applications of precursory EM signals in real time EQ prediction continue to be challenging due to several problems; (i) intensity of anticipated seismo-EM signals in ULF band is very low (notable exception being the highly enhanced signals recorded in association with M7.1 Loma Prieta earthquake (Fraser-Smith et al., 1990; Bleier et al., 2009), although question has been raised that this anomalously large signal may not be due the proximity of the magnetometer to the epicenter but artifact of the sensor-system malfunction (Thomas et al., 2009)), (ii) discrimination of weak seismo-EM signals from the background natural EM fields of ionospheric and magnetospheric origin and (iii) finally the limitation in the localization of precursor source or, at least, determination of azimuth direction to the source zone. Very often later problems are aggravated by short time (less than 5 min) of precursor existence (Bleier et al., 2009). With the availability of very sensitive induction type 3-component magnetometers with high suppression of man-made interference, the recording of high quality magnetic data in ULF bands has greatly improved (Hayakawa et al., 2007). For the second problem, polarization analysis incorporating the ratio S_z/S_H (S_z and S_H are the spectral intensities of vertical and horizontal magnetic field components) is found effective, at least partially, in distinguishing seismo-EM signals from geomagnetic field fluctuations (Hayakawa et al., 1996). The formulations of principal component analysis and fractal approach have been used with some success in isolating components of extra-terrestrial and seismotectonic origin in magnetic field records (see, for example, Hayakawa et al., 1999, 2007; Serita et al., 2005; Ida and Hayakawa, 2006). Towards the identification of underground ULF source or its direction, formulation based on the time lag or phase difference between pair or more observation points has been advanced, referred to as gradiometric method (Kopytenko et al., 2001, 2006; Ismaguilov et al., 2003). The technique may lack reliability as the ULF electromagnetic waves propagating through the conductive layers undergo strong dissipation and dispersions, making the identification of front of ULF signal ambiguous (Surkov et al., 2004). Working independently, Surkov et al. (2004) advocated use of amplitude difference in synchronous observation at two or multiple recording sites. The method, in principle, hold promise to estimate both the location and direction of the ULF source provided the spatial scale of natural noise variation should be much greater than both characteristic length of the ULF signal itself and distance between magnetometers. However, space derivative of magnetic field perturbations tend to be very unstable at low signal-to-noise (S/N) ratio and gives a big error in the estimation of source direction (see remarks about S/N ratio in Dudkin et al., 2003). Very promising in the direction-finding problem for seismo- EM precursors is an application of the polarization ellipse (PE) technique, where the PE major axis behavior is investigated (goniometric method) (Du et al., 2002; Schekotov et al., 2007, 2008). This technique allows determination of trends in azimuth angle of anomalous ULF signal and possibly area of EQ epicentre. Taking into account that ULF magnetic source is always in the PE plane the new method of magnetic precursor source location from two observation points has been proposed by present authors (Dudkin et al., 2008). In the present paper expanding on the steps of this new direction-finding approach, we test apply the formulation on magnetic field data from pair of stations operated simultaneously in one of the seismoactive region of India.

The paper is structured as follow: Sect. 2 gives brief account of the seismic background in the study area. The noisesensitivity parameters of magnetometer deployed, data acquisition and initial data preparation steps and nature of the seismic activity recorded during the observational campaign are listed in Sect. 3. The outline of the PE method for the isolation of seismo-EM signals and estimation of the direction and possible location of seismo-EM source is described in Sect. 4. In Sect. 5, discussion of the results in reference to source validation and possible generating source mechanism form the central stage, followed by a short summary.

CURRENT AND PAST SEISMIC HISTORY OF THE STUDY AREA AND EXPERIMENT DESCRIPTION

Koyna-Warna region in the southern part of Deccan Volcanic Province, Western India is a classical example of reservoir triggered seismicity – Fig. 1 (for review see Gupta, 2002, and references therein). The largest triggered earthquake of M 6.3 occurred here on 10 December 1967, and over the past four and a half decades area has remained intensively seismoactive (Gupta, 2005). During this period 19 EQs of $M_L \geq 5$ and about 170 EQs with $M_L \geq 4$ have occurred, all confined to a well-defined belt of roughly 20×30 km^2 with hypocentres h $\leq$ 12 km (Gupta et al., 2007). Such features make the area unique for studying peculiarities of magnetic field during EQ preparation process.

To develop and test the PE method for locating source region of EM fields produced during EQ preparation process, two stations to record magnetic variations in ULF bands were established; the first at Koyna, within the limits of focused seismic zone, and other was placed at distant location of ~100 km in Shivaji University, Kolhapur (Fig. 1). Both sites were located in low-noise background and have minimal interference from man-made disturbances.

Extremely low-noise ($S_{B,n}$=0.2 pT/Hz$^{0.5}$ at f =1 Hz, where $S_{B,n}$ is spectral noise density) with high factor of industrial interference suppression (more than 1000) 3- component LEMI-30 magnetometers, specially designed by Lviv Centre of Institute for Space Research, Ukraine (http: //www.isr.lviv.ua) for EQ EM monitoring, were deployed at both stations for synchronous recording. LEMI-30 magnetometers operate in frequency range 0.001–32 Hz and are ideally suited to record the most promising EQ magnetic precursors in ULF band, which are found dominant below 0.1 Hz (Hayakawa et al., 2004, 2007).

Both stations recorded simultaneous data with 64 samples per second over the entire observational campaign period of 15 March–30 June 2006. During data processing a resampling procedure was applied with averaging each of 64 samples. Thus an upper boundary of signal spectra was decreased to 0.5 Hz. For such data the dynamical Fourier spectra (DFS) for each 24 h of data recording were calculated. Then for each point of DFS the parameters of PE for each measuring site were calculated, which form the base to search EQ precursory magnetic analysis described in the next section.

In the post 1993 period, the seismic activity in Koyna- Warna region is being monitored by a closely spaced network of seven modern 3-component seismometers established by National Geophysical Research Institute, Hyderabad (Gupta, 2002). The uniform azimuth coverage allows estimation of epicentres to an accuracy of ±0.5 km. During the period of present campaign, more than 700 EQs with magnitude in the range of M_L=0.5–4.7 were recorded, two of them with magnitude M_L >4, 8 EQs with ML >3 and as many as 172 with M_L >2. The spatial distribution of these EQs with M_L >2.5 (Fig. 1) completely overlaps with the area outlined by seismic activity over the past 45 years. The 2 largest EQs occurred; EQ1 on 17 April 2006 (M_L=4.7, h=3.9 km, 17.13 N, 73.78 E, 16.39.59.4 UT) and EQ2 on 21 May 2006 (M_L=4.2, h=5.1 km, 17.17 N, 73.77 E, 20.29.01.2 UT) and their fault plane solutions as determined by NGRI are shown as inset in Fig. 1. The ULF magnetic activity in relation to these 2 modest magnitude EQs is examined and presented in next sections.

EXPERIMENT, RESULTS AND DISCUSSION

Polarization Analysis

As described earlier, the polarization analysis in the form of S_Z/S_H plots in the select frequency bands are commonly used to search precursory ULF magnetic signals before large earthquakes. Figure 2 gives the plot of S_Z/S_H at a representative frequency band of 0.01–0.03 Hz (Yumoto et al., 2009) for the April–May 2006. This frequency band is selected as it is shown later that most precursory magnetic signals in ULF bands are concentrated in frequency range of 0.01–0.07 Hz. This simple mode of presentation shows that except for oc- casional differences in the temporal variability, polarization ratios at both stations exhibit general resemblance. However, the variability of S_Z/S_H ratios neither show any significant correlation with geomagnetic activity nor any significant change from background values, which could be classified as magnetic precursor candidates (see Fig. 2), is seen in relation to EQ occurrence. Given the high sensitivity of the LEMI-30 search coil magnetometer used here, magnetic signals with amplitude as low as 1–80 pT on H and Z are well resolved but it seems likely that small amplitude EQ precursory magnetic signals, even if present in relation to the moderate magnitude EQ studied here, are completely masked by much stronger signals from ionospheric and magnetospheric sources to stand out clearly in S_Z/S_H ratio plot.

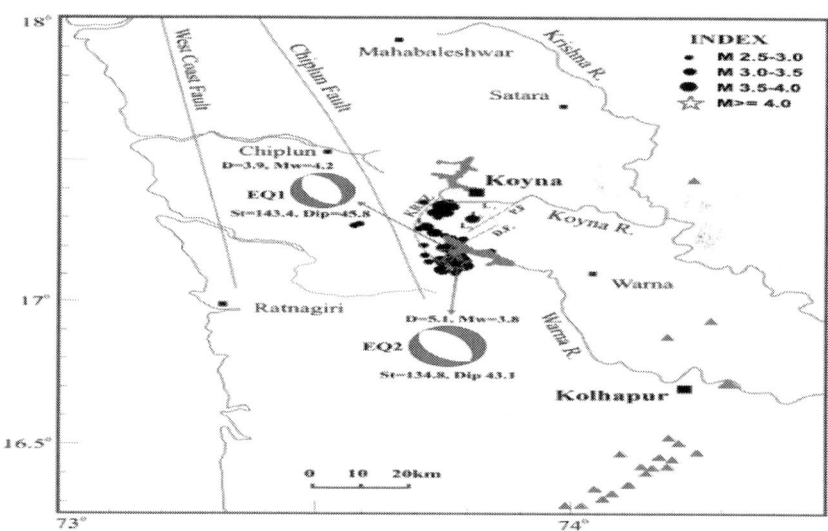

Figure 1: Map of western India showing location of ULF magnetometer sta- tions established to monitor EM emission related to reservoir triggered seis- micity associated with Koyna-Warna reservoirs. Black dots represent epicen- tre of EQs recorded during the observational campaign period of March–June 2006. Fault plane solutions of two moderate EQs (M_L >4) are also shown. Surface intersection point of M-lines (see text) are indicated by red and blue triangles.

Source Azimuth Estimation by Polarization Ellipse Method

To extract information on the seismic source from the directional dependence of the ULF magnetic field components, the following data processing steps were carried out under some simplifying assumptions:

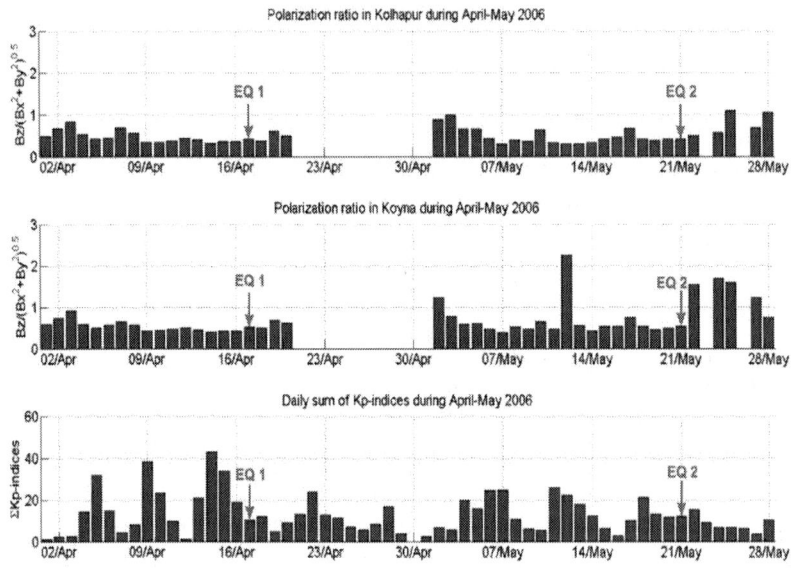

Figure 2: Polarization ratio S_Z/S_H in frequency range 0.01–0.03 Hz and Kp sums during March–June 2006. The gaps in data relate to days of mainpower disconnections.

- In narrow ULF frequency band the magnetic field components can be represented as harmonic (periodic) signals.

- At a given frequency, 3-orthogonal components of magnetic field with specific phases provide estimate the PE parameters, the resultant PE plane in space contains the source of EM field.

- Alternating (induction) currents generated by any alternative seismo-EM source mechanisms, referred to in Sect. 1, can be visualized as small-scale superposition on large-scale telluric current system induced by global induction. These local current perturbations resulting from the motion of the conductive layer of rocks (inductive effects) or related to piezoelectric/ piezomagnetic effects or so can generally be viewed as closed loop configuration and thus can be approximated to be equivalent to that produced by elementary magnetic dipoles placed in the source region. The assumption of magnetic-dipole type configuration is implicit in estimating the ULF source parameters using the spatial derivative of magnetic field (Surkov et al., 2004). In case of electrokinetic effect, the superposed current system may contain unclosed configuration (e.g. see Fedorov et al., 2001) and thus inhibit approximation of net flow by magnetic-type dipole. This would especially be valid when flow is confined to uniform homogeneous medium (Moore et al., 2004). However, for real crust characterized by inhomogeneous structures, it has been shown that when the underground water under the effect of accumulating stresses is forced through the narrow fault plane the resulting streaming potential gives rise to concentrated flow of electric current along the fault plane that closes its path by way of return currents on either side of the fault plane (Mizutani and Ishido, 1976). This is consistent with the generalized calculations of Moore et al. (2004) who showed that net current resulting from electrokinetic effect consists of "impressed" and "back" (return) currents. In such a special structural geometry, the overall configuration of electrokinetic current flow can form compact asymmetric closed loops with a component of intense current along highly conductive duct (fault) and thus allowing magnetic dipole-like source approximation. Such approximation appears justified for the present study region where source region of seismicity is confined to short length (~20 km) of narrow Koyna River Fault Zone (Gupta et al., 2007). This fault segment form steep boundary between inhomogeneous crustal blocks and provide conduits for fluid pressure flow which prevails right up to hypocenter depths (Talwani, 1997a; Pandey, 2003; Agrawal et al., 2004).

- Magnetic moment of such a magnetic dipole source is in PE plane formed by its field components at the measurement points.

So far as the PE plane at any time contains the source of magnetic field, it is possible to find the intersection line of PE planes from observations at two distant stations (Fig. 3). This line, which we name as M-line, contains the magnetic dipole moment M, which is aligned along it, and can be calculated from parameters of PE. Calculation of PE parameters and magnetic moment follows the details can be found in relevant monographs on electromagnetism or in article (Morgan and Evans, 1951).

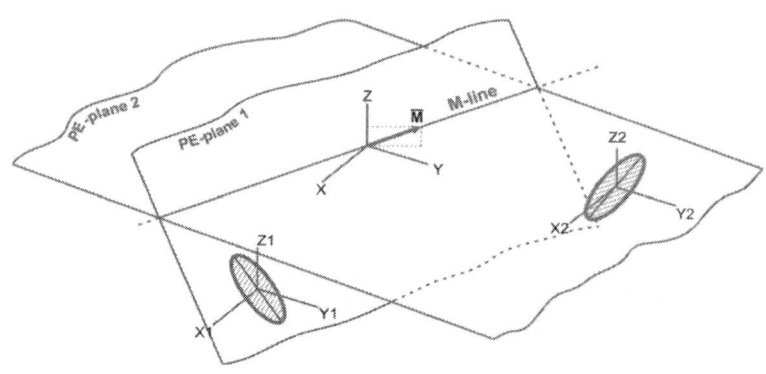

Figure 3: Formation of M-line by the intersection of two PE-planes.

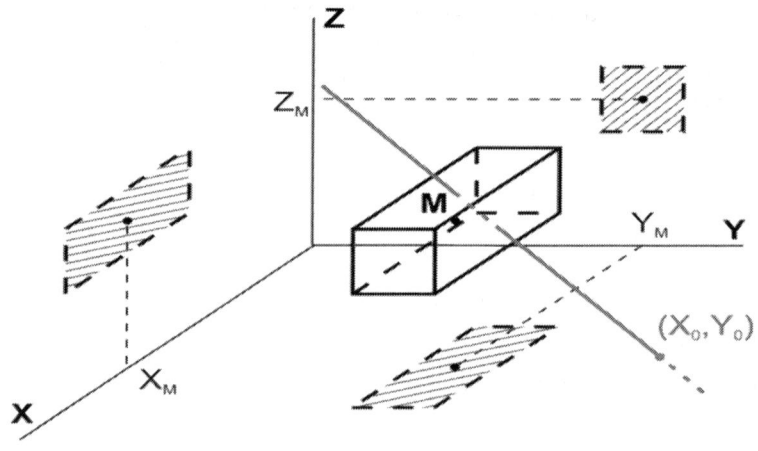

Figure 4: Configuration of M-area.

Source region of the EQs near Koyna is confined to narrow localized area 20×30 km² with hypocentres restricted to depths less than 12 km (Gupta, 2005; Gupta et al., 2007). Thus for the detection of the magnetic precursors of two main EQs, we defined the source area of interest (M-area) by a parallelepiped of 20 km length and a cross section of 3×3 km² with centre at EQ hypocentres. Then we calculate PE planes intersections (M-lines) which intersect M-area (Fig. 4) for both stations of synchronous observations. The M-lines were calculated for each point of DFS in the frequency range 0.001–0.5 Hz (i.e. for each harmonic and elementary time window used in the calculation of spectra).

It is widely accepted that the most promising frequency range of ULF magnetic precursors overlaps with the Pc3- Pc5 band of micropulsations (see, for example, Hayakawa et al., 2007; Kopytenko et al., 2001; Molchanov et al., 2004). The sources of these pulsations are dominantly of magnetospheric origin related to field line resonance or large-scale Figure. 5. Koyna/Kolhapur PE major axes ratio against direction of the magnetic dipole cavity oscillations (see for example Waters et al., 1994). The weak secondary ionospheric component arises due to instantaneous penetration of the magnetospheric inducted polar electric field to low and middle latitude (Trivedi et al., 1997). The spatial scale-length of these magnetospheric/ionospheric current system at low and middle latitudes are typically several thousand kilometers and, therefore, permits plane wave approximation for natural electromagnetic waves (Cagniard, 1953). A net consequence of the plane wave approximation on the present analysis is that the ratio of the amplitude of magnetospheric ULF magnetic signals at pair of measuring stations, separated by not more than couple of hundreds of km, would be close to unity. In contrast, characteristics spatial scale-length of the ULF signal is no more than few hundred km (Hayakawa et al., 2007). Taking advantage of these large differences in dimension of the two sources, for the discrimination of M-lines associated with seismo-EM sources from natural background variations, we calculated the ratio of PE major axes in Koyna and Kolhapur measuring sites against orientation of horizontal magnetic dipole placed in EQ hypocentres by simplified formulas, The results of calculation are shown on Fig. 5, where the minimal ratio of PE major axes is about 2. Therefore, it appears that allowing for contamination for background and man-made noise, the M-lines with PE major axes ratio exceeding the threshold value 2 can be ascribed to the ULF magnetic precursor candidate.

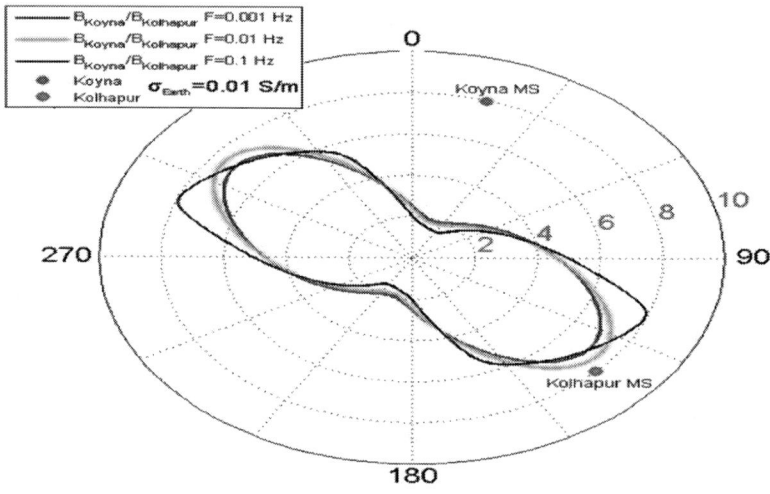

Figure 5: Koyna/Kolhapur PE major axes ratio against direction of the magnetic dipole moment.

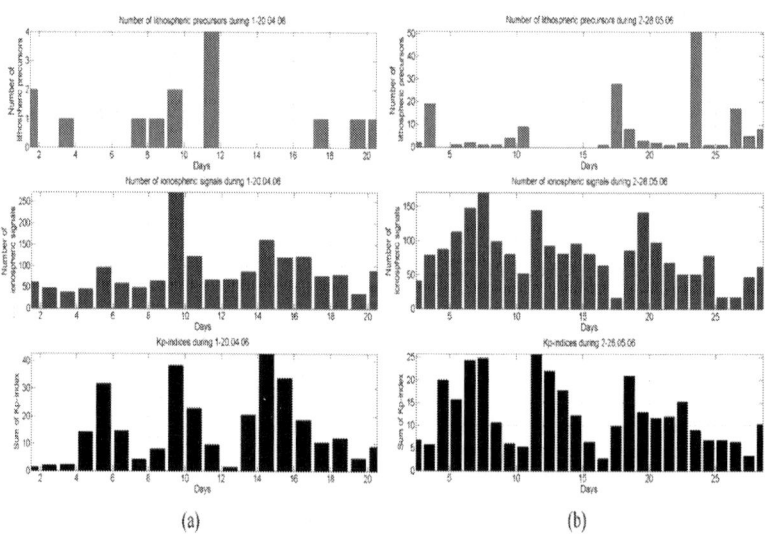

Figure 6: The number of detected seismo-EM and ionospheric signals with the Kp-index values during time of observation. Red bars are seismo-EM precursors, blue bars are ionospheric signals.

Source Validation

The detected and classified seismo-EM and ionospheric signals with the Kp-index values during the observation campaign interval are shown in Fig. 6a, b. The good correlation between the numbers of signals of the ionospheric origin and the value of Kp-index is clearly seen. The number of seismo-EM signals increase before EQ1 (M_L=4.7) up to 11 April and then approaches to zero level. After EQ1 very low seismo-EM activity in region of interest is observed. For EQ2 (M_L=4.2) the number of magnetic precursors is maximal on 17 May and then drop rapidly to small value. Then on 23 May the signals classified as seismo-EM origin rise again, which apparently is neither related to any abnormal seismic activity nor marked by intense solar/magnetic disturbances. However, given that this time interval is marked by moderate values of Kp, it seems possible that these seismo-EM signals related to the release of residual mechanical stresses following EQ2.

Further as an example, the frequency of the seismo- EM and ionospheric/magnetospheric signals against time are shown for 17, 18 and 20 May (see Fig. 7). The magnetic precursors are confined to narrow frequency range of about 0.01–0.07Hz and completely overlap with the most dominant frequency range of ionospheric signals. Thus, source separation based on the frequency characteristic alone will not be unambiguous. However, it is shown that the ratio of major axes of PE at two distant sites proves effective in isolating EM signals of seismo-EM and inospheric/ magnetospheric origins (Figs. 6 and 7). It may be added that fixing of major axes ratio at 2 for isolating signals of different origin is some what arbitrary. Adopting value below or above the critical threshold would, respectively, lead to either increased cases of contamination from ionospheric/magnetospheric sources or reduced number of precursory events for the purpose of further estimation of Mlines aimed at primary goal of source location. May be long term monitoring of magnetic fields in ULF band in a given seismic belt would help more realistic identification of factor.

EM Source Mechanism and Linkage to Seismicity

In the present study, M-lines estimations were done for all spectral lines by fixing major axes ratio at 2. The cluster of points resulting from the intersection of calculated M-lines with the earth surface are superposed in Fig. 1 for all those spectral lines which are identified as possible magnetic precursor to the EQ2 (red dots in Fig. 7). It follows from the definition of M-lines, the line joining surface intersection points to the epicentral zone gives the azimuth of magnetic ULF source and similarly line connecting these surface interaction points with hypocentral zone gives the ascent angle of the source magnetic moment. It follows from Fig. 1 that azimuth of potential magnetic ULF precursors orient primarily in NNW-SSE direction. This direction closely follows the orientation of major lineament, mega fracture and geomorphic (fissure) features of the region (Talwani, 1997a). It is known that prolonged seismic activity in the Koyna-Warna region starting from 1963 to the present has concentrated on a short 30 km long segment of the fault, which is seen as an extension of 10° N–15° E trending Koyna River Fault Zone (KRFZ) to the south in NNW-SSE direction (Fig. 1).

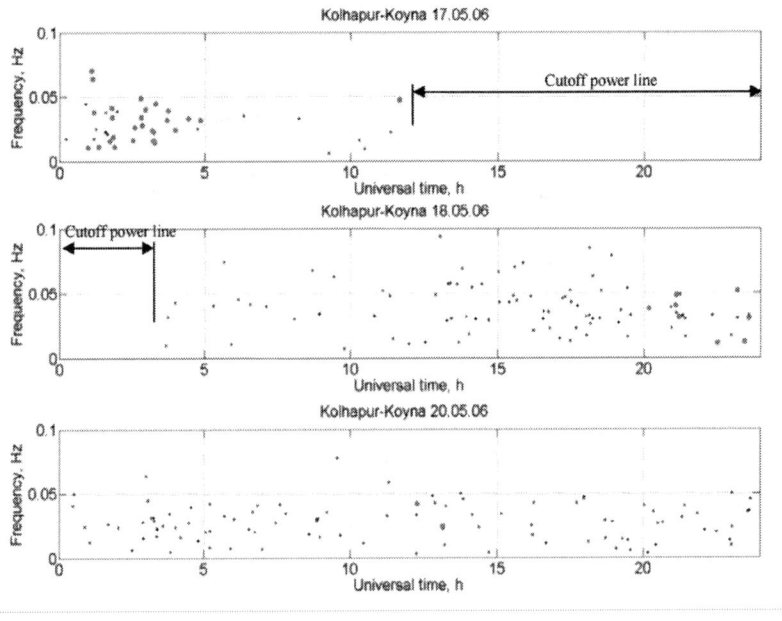

Figure 7: Distribution of ionospheric and seismo-EM M-lines against time and frequency for 17, 18 and 20 May 2006. Red and blue points are seismo-EM and ionospheric signals respectively.

Another notable feature of the seismicity in this region is that epicentres of all large EQs (M>5) starting with the Koyna mainshock in 1967 (M=6.3) to last EQ of M>5 in 2005 have progressively migrated southward on this NNW-SSE trending fault starting from the south of the Koyna river and extending to the Warna River (Talwani, 1997b). Amongst the several mechanisms suggested to explain the occurrence of continued seismicity in the Koyna-Warna region, the role of high-fluid pressure in producing failure on the preexisting critically stressed faults has received wider acceptance in generating EQs in the region (Talwani, 1997a; Pandey, 2003). More recently, extending this hypothesis, it have been attributed the temporal southward migratory trend in seismicity to similar changes in direction of pore-fluid pressure diffusion (Ramalingeswara and Singh, 2008). They further demonstrated that changes in fluid flow directions were independently corroborated by the strikes of the focal mechanisms of large EQs (M>5). The fault plane solution of EQ1 and EQ2 indicate normal faulting, in agreement with the faults mechanisms for most of the large EQs recorded from this region (Gupta et al., 2002). Further, one of strike directions of fault plane solutions of the EQ1 and EQ2 is parallel to the active segment of the KRFZ. Given this correspondence and observational evidence that azimuth plane of magnetic ULF sources is oriented in the same direction, it is possible to postulate that both EQs and ULF magnetic signals are common manifestation of fluid flow diffusion in the focal zone. The active segment of the KRFZ, multiple fractures and fissure may be providing conduits for the high pore-pressure flow down to hypocentral depth (Talwani, 1997b; Agrawal et al., 2004). The oriented focused flow of fluid at focal depth under the effect of accumulated high stresses by way of electrokinetic effects may be the effective source cause for ULF magnetic signals (Mizutani et al., 1976; Uyeda et al., 2009, and references therein). The electrokinetic effect was earlier invoked by Arora (1988; also see Arora and Singh, 1992) to explain the bi-polarity nature of seismomagnetic anomalies observed around the Koyna reservoir in association with a sequence of moderate magnitude (M_L=4.1–4.3) EQs in the same area during 1980.

SUMMARY

The use of highly sensitive and low noise search coil LEMI- 30 magnetometers enabled to resolve ULF magnetic signals in frequency range of 0.001–0.5 Hz with amplitudes as lowas of few pT. Despite this advantage, the commonly employed polarization analysis (S_z/S_H ratio) fails to reveal any precursory seismo-EM signals, perhaps such signals, even if present, are completely masked by the much stronger signals of natural ionospheric/magnetospheric origin. This is rather not surprising as almost all reported anomalies are in relation to large EQs of M>5, whereas the only two EQs searched here have magnitude M_L >4. However, controlled by the orientation of seismogenic faults, resulting seismo-EM field would have definite orientation in comparison to the isotropic direction distribution of highly variable natural signals arising from complex ionospheric-magnetospheric interactions. Based on these physical considerations, the interactions lines defined by the planes of PE, formed by the magnetic fields at minimum two sites, define the azimuth of seismo-EM source. Further, ratio of major axes of PEs above certain threshold help to distinguish ULF signals dominated by seismo-EM origin from those associated with ionospheric origin. In the present case, this threshold is fixed at 2, corresponding to the minimum value of the ratio recorded between Koyna and Kolhapur. This choice of the value of cutoff threshold adopted here has arbitrariness but can be constrained better with long term data. Approximating the seismo-EM source as elementary magnetic dipole, the large numbers of spectral lines qualifying this threshold provide statistically averaged azimuth of the seismo-EM field. The NNW-SSE orientation of seismogenic ULF signals in the Koyna-Warna corresponds well with causative fault zone inferred from longterm EQ data. The already available knowledge on the role of high pressure fluids in generating the EQs favours electrokinetic effect to be one of the possible source mechanisms for seismo-EM fields. The alternative source mechanisms, inductive or piezomagnetic effects, may equally well explain the observations. Testing the proposed formulation to the other more active seismic belts, where source zone is not as well defined as the Koyna-Warna region would help generalization of the methodology for future EQ precursory studies. Such future experimentation essentially employing multiple stations would be key diagnostic to map the exact nature of sesimo-EM field perturbations and thus validating the magnetic dipole-like approximation for the source, implicit to the formulation advanced here.

ACKNOWLEDGEMENTS

At Wadia Institute of Himalayan Geology, the work presented above was supported by the Ministry of Earth Sciences, Government of India under a sponsored project "Setting up of Multi-Parametric Geophysical Observatories for Earthquake Precursory Research", while part of the analysis at Lviv Centre was partially supported by the STCU grants 3165. The magnetometer campaign, described here, was initiated on the advice of Harsh K. Gupta, Chairman of the Program Advisory Group. Many formal discussions with him about the Koyna seismicityand guidance are greatly acknowledged. B. K. Bansal, Program Co-ordinator, MoES is thanked for facilitating the activity and suggestions in experiment planning. Authors have special thanks to D. Shashidhar, NGRI, Hyderabad for readily providing the EQ data including the fault plane solutions. The help received from Shri D. D. Khandelwal in field campaign is deeply appreciated. Authors thank the parent institutions for providing facilities in bringing the present study to this stage.

REFERENCES

1.   Agrawal, P. K., Pandey, O. P., and Chetty, T. R. K.: Aeromagnetic anomalies, lineaments and seismicity in Koyna-Warna Region, J. Ind. Geophys. Union, 8(4), 229–242, 2004.

2.   Arora, B. R.: Tectonomagnetic studies in India (Invited), "Earthquake Prediction – Present Status", edited by: Guha, S. K. and Patwardhan, A. M., Dept. of Geology, University of Poona, Pune, India, pp. 53–62, 1988.

3.   Arora, B. R. and Singh, B. P.: Geomagnetic and geoelectric investigations for seismicity and seismotectonics of the Himalayan region, in: Himalayan Seismicity, edited by: Gupta, G. D., Mem. Geol. Soc. India., 23, 223–262, 1992.

4.   Ba ̃nos, A.: Dipole radiation in the presence of a conducting halfspace, Pergamon Press, Oxford, 1966.

5.   Bleier, T., Dunson, C., Maniscalco, M., Bryant, N., Bambery, R., and Freund, F.: Investigation of ULF magnetic pulsations, air conductivity changes, and infra red signatures associated with the 30 October Alum Rock M5.4 earthquake, Nat. Hazards Earth Syst. Sci., 9, 585–603, doi:10.5194/nhess-9-585-2009, 2009.

6.   Cagniard, L.: Basic theory of the magnetotelluric method of geophysical prospecting, Geophysics, 18, 605–635, 1953.

7.   Du, A., Huang, Q., and Yang, S.: Epicenter location by abnormal ULF electromagnetic emissions, Geophys. Res. Lett., 29(10), 1455–1458, 2002.

8.   Dudkin, F., De Santis, A., and Korepanov, V.: Active EM sounding for early warning of earthquakes and volcanic eruptions, Phys. Earth Planet. Inter., 139(3, 4), 187–195, 2003.

9.   Dudkin, F., Leontyeva, O., Arora, B. R., Rawat, G., and Sharma, A.: Analysis of magnetic field polarization parameters before and after Koyna earthquakes, EGU General Assembly, Vienna, Austria, 13–18 April 2008, Geophys. Res. Abstr., 10, EGU2008-A-00054, 2008.

10.   Fedorov, E., Pilipenko, V., and Uyeda, S.: Electric and Magnetic Fields Generated by Electrokinetic Processes in a Conductive Crust, Phys. Chem. Earth (C), 26(10–12), 793–799, 200l.

11.   Fitterman, D. V.: Theory of electrokinetic magnetic anomalies in a faulted half-space, J. Geophys. Res., 84(B11), 6031–6040, 1979.

12.   Fraser-Smith, A. C., Bernardi, A., McGill, P. R., Ladd, M. E., Helliwell, R. A., and Villard Jr., O. G.: Low-frequency magnetic field measurements near the epicenter of the Ms 7.1 Loma Prieta earthquake; Geophys. Res. Lett., 17, 1465–1468, 1990.

13.   Gupta, H. K.: A review of recent studies of triggered earthquakes by artificial water reservoirs with special emphasis on earthquakes in Koyna, India, Earth-Sci. Rev., 58, 279–310, 2002.

14.   Gupta, H. K.: Artificial water reservoir-triggered earthquakes with special emphasis at Koyna, Current Science, 88(1), 1628–1631, 2005.

15.   Gupta, H., Shashidhar, D., Pereira, M., Mandal, P., Purnachandra, Rao N., Kousalya, M., Satyanarayana, H. V. S., and Dimri, V. P.: Earthquake forecast appears feasible at Koyna, India, Current Sci., 93(6), 1628–1631, 2007.

16.   Hayakawa,M., Kawate, R., Molchanov, O. A., and Yumoto, K.: Results of ultra-low-frequency magnetic field measurements during the Guam earthquake of 8 August 1993, Geophys. Res. Lett., 23, 241–244, 1996.

17. Hayakawa, M., Itoh, T., and Smirnova, N.: Fractal analysis of ULF geomagnetic data associated with the Guam earthquake on 8 August 1993, Geophys. Res. Lett., 26, 2797–2800, 1999.

18. Hayakawa, M., Itoh, T., Hattori, K., and Yumoto, K.: ULF electromagnetic precursors for an earthquake in Biak, Indonesia on 17 February 1966, Geophys. Res. Lett., 27, 1531–1534, 2000.

19. Hayakawa, M., Molchanov, O. A., and NASDA/UEC team: Achievements of NASDA's Earthquake Remote Sensing Frontier Project, TAO, 15(3), 311–327, 2004.

20. Hayakawa, M., Hattori, K., and Ohta, K.: Monitoring of ULF (ultralow- frequency) Geomagnetic Variations Associated with Earthquakes, Sensors, 7, 1108–1122, 2007.

21. Ida, Y. and Hayakawa, M.: Fractal analysis for the ULF data during the 1993 Guam earthquake to study prefracture criticality, Nonlin. Processes Geophys., 13, 409–412, doi:10.5194/npg-13-409-2006, 2006.

22. Ismaguilov, V. S., Kopytenko, Yu. A., Hattori, K., and Hayakawa, M.: Variations of phase velocity and gradient values of ULF geomagnetic disturbances connected with the Izu strong earthquakes, Nat. Hazards Earth Syst. Sci., 3, 211–215, doi:10.5194/nhess-3-211-2003, 2003.

23. Johnston, M. J. S., Muller, J. S., and Sasai, Y.: Magnetic field observations in the near field: the 28 June, 1992 Mw 7.3 Landers, California Earthquake, B. Seism. Soc. Am., 84, 792–798, 1994.

24. Kopytenko, Yu. A., Ismaguilov, V. S., Hayakawa, M., Smirnova, N., Troyan, V., and Peterson, T.: Investigation of the ULF electromagnetic phenomena related to earthquakes: contemporary achievements and perspectives, Annali di Geofisica, 44(2), 325– 334, 2001.

25. Kopytenko, Yu. A., Ismaguilov, V. S., Hattory, K., and Hayakawa, M.: Determination of hearth position of forthcoming strong EQ using gradients and phase velocities of ULF geomagnetic disturbances, Phys. Chem. Earth, 31, 292–298, 2006.

26. Martin, R. J., Habermann, R. E., and Wyss, M.: The effect of stress cycling and inelastic volumetric strain on remanent magnetization, J. Geophys. Res., 83, 3485–3496, 1978.

27. Mizutani, H. and Ishido, T.: A new interpretation of the magnetic field variation associated with the Matsushiro Earthquakes, J. Geomagn. Geoelectr., 28, 179–188, 1976.

28.   Mizutani, H., Ishido, T., Yokokura, T., and Ohnishi, S.: Electrokinetic phenomena associated with earthquakes, Geophys. Res. Lett., 13, 365–368, 1976.

29.   Molchanov, O. A., Kopytenko, Yu. A., Voronov, P. M., Kopytenko, E. A., Matiashvili, T. G., Fraser-Smith, A. C., and Bernardy, A.: Results of ULF magnetic field measurements near the epicenters of the Spitak (Ms = 6.9) and Loma Prieta (Ms = 7.1) earthquakes: comparative analysis, Geophys. Res. Lett., 19, 1495–1498, 1992.

30.   Molchanov, O. A. and Hayakawa, M.: Generation of ULF electromagnetic emissions by microfracturing, Geohpys. Res. Lett., 22, 3091–3094, 1995.

31.   Molchanov, O. A., Schekotov, A. Yu., Fedorov, E., Belyaev, G. G., Solovieva, M. S., and Hayakawa, M.: Preseismic ULF effect and possible interpretation, Ann. Geophys., 47(1), 119–131, 2004.

32.   Moore, J. R., Glaser, S. D., Morrison, H. F., and Hoversten, G. M.: The streaming potential of liquid carbon dioxide in Berea sandstone, Geophys. Res. Lett., 31, L17610, doi:10.1029/2004GL020774, 2004.

33.   Morgan, M. and Evans, W.: Synthesis and analysis of elliptic polarization loci in terms of space-quadrature sinusoidal components, Proc. IRE, 39, 552–556, 1951.

34.   Ogawa, T., Oike, K., and Miura, T.: Electromagnetic Radiations from Rocks, J. Geophys. Res., 90(D4), 6245–6249, 1985. Ogawa, T. and Utada, H.: Coseismic piezoelectric effects due to a dislocation. 1. An analytic far and early-time field solution in a homogeneous whole space, Phys. Earth Planet. Inter., 121, 273– 288, 2000.

35.   Pandey, A. P. and Chadha, R. K.: Surface loading and triggered earthquakes in the Koyna–Warna region, western India, Phys. Earth Planet. Inter., 139, 207–223, 2003.

36.   Ramalingeswara Rao, B. and Singh, C.: Temporal migration of earthquakes in Koyna–Warna (India) region by pore-fluid diffusion, J. Seismol., 12(4), 547–556, 2008.

37.   Schekotov, A. Y., Molchanov, O. A., Hayakawa, M., Fedorov, E. N., Chebrov, V. N., Sinitsin, V. I., Gordeev, E. E., Belyaev, G. G., and Yagova, N. V.: ULF/ELF magnetic field variations from atmosphere induced by seismicity, Radio Sci., 42, RS6S90, doi:10.1029/2005RS003441, 2007.

38. Schekotov, A. Y., Molchanov, O. A., Hayakawa, M., Fedorov, E. N., Chebrov, V. N., Sinitsin, V. I., Gordeev, E. E., Andreevsky, S. E., Belyaev, G. G., Yagova, N. V., Gladishev, V. A., and Baransky, L. N.: About possibility to locate an EQ epicenter using parameters of ELF/ULF preseismic emission, Nat. Hazards Earth Syst. Sci., 8, 1237–1242, doi:10.5194/nhess-8-1237-2008, 2008.

39. Serita, A., Hattori, K., Yoshino, C., Hayakawa, M., and Isezaki, N.: Principal component analysis and singular spectrum analysis of ULF geomagnetic data associated with earthquakes, Nat. Hazards Earth Syst. Sci., 5, 685–689, doi:10.5194/nhess-5-685-2005, 2005.

40. Surkov, V. V.: ULF electromagnetic perturbations resulting from the fracture and dilatancy in the earthquake preparation zone, pp. 357–370, in: Atmospheric and Ionospheric Phenomena Associated with Earthquakes, edited by: Hayakawa, M., TERRAPUB, Tokyo, 1999.

41. Surkov, V. V., Molchanov, O. A., and Hayakawa, M.: Preearthquake ULF electromagnetic perturbations as a result of inductive seismomagnetic phenomena during microfracturing, J. Atmos. Sol. Terr. Phys., 65(1), 31–46, 2003.

42. Surkov, V. V., Molchanov, O. A., and Hayakawa, M.: A direction finding technique for the ULF electromagnetic source, Nat. Hazards Earth Syst. Sci., 4, 513–517, doi:10.5194/nhess-4-513-2004, 2004.

43. Talwani, P.: On the nature of reservoir-induced seismicity, Pure Appl. Geophys., 150, 473–492, 1997a.

44. Talwani, P.: Seismotectonics of the Koyna-Warna Area, India, Pure Appl. Geophys., 150, 511–550, 1997b.

45. Thomas, J. N., Love, J. J., and Johnston, M. J. S.: On the reported magnetic precursor of the 1989 Loma Prieta earthquake, Phys. Earth Planet. Int., 173, 207–215, 2009.

46. Trivedi, N. B., Arora, B. R., Padilha, A. L., Da Costa, J. M., and Dutra, S. L. G.: Global Pc5 geomagnetic pulsations of March 24, 1991, as observed along the American sector, Geophys. Res. Lett., 24, 1683–1686, 1997.

47. Uyeda, S., Nagao, T., and Kamogava, M.: Short-term earthquake prediction: Current status of seismo-electromagnetics, Tectonophysics, 470, 205–213, 2009.

48. Yumoto, K., Ikemoto, S., Cardinal, M. G., Hayakawa, M., Hattori, K., Liu, J. Y., Saroso, S., Ruhimat, M., Husni, M., Widarto, D., Ramos, E., McNamara, D., Otadoy, R. E., Yumul, G., Ebora, R., and Servando, N.: A new ULF wave analysis for Seismo-Electromagnetics using CPMN/MAGDAS data, Phys. Chem. Earth, 34, 360–366, 2009.

49. Waters, C. L., Menk, F.W., and Fraser, B. J.: Low latitude geomagnetic field line resonance: Experiment and modeling, J. Geophys. Res., 99(A9), 17547–17558, 1994.

2

Phenomena of Electrostatic Perturbations before Strong Earthquakes (2005–2010) Observed on DEMETER

X. Zhang[1], X. Shen[1], M. Parrot[2], Z. Zeren[1], X. Ouyang[1], J. Liu[1], J. Qian[1], S. Zhao[1], and Y. Miao[1]

[1]Institute of Earthquake Science, CEA, Beijing 100036, China
[2]LPC2E/CNRS, 3A Avenue de la Recherche Scientifique, 45071 Orl´eans cedex 2, France

ABSTRACT

During the DEMETER operating period in 2004–2010, many strong earthquakes took place in the world.69 strong earthquakes with a magnitude above 7.0 duringJanuary 2005 to February 2010 were collected and analysed.The orbits, recorded in local nighttime by satellite,were chosen by a distance of 2000 km to the epicentres duringthe 9 days around these earthquakes, with 7 days beforeand 1 day after. The anomaly is defined when the disturbancesin the electric field PSD increased to at least 1 orderof magnitude relative to the normal median level about$10^{-2}\mu V^2/m^2/Hz$ at 19.5–250 Hz frequency band, and thestarting point of perturbations not exceeding 10° relativeto the epicentral latitude. Among the 69 earthquakes, it isshown that electrostatic perturbations were detected at ULFultralow frequency and ELF-extremely low frequency bandbefore the 32 earthquakes, nearly 46 %. Furthermore, weextended the searching scale of these perturbations to theglobe, and it can be found that before some earthquakes,the electrostatic anomalies were distributed in a much largerarea a few days before, and then they concentrated to theclosest orbit when the earthquake would happen one dayor a few hours later, which reflects the spatial developingfeature during the seismic preparation process. The resultsin this paper contribute to a better description of theelectromagnetic (EM) disturbances at an altitude of 660–710 km in the ionosphere that can help towards a further understandingof the lithosphere-atmosphere-ionosphere (LAI) coupling mechanism.

INTRODUCTION

Spatial electromagnetic phenomena have been widely observed by satellites, including the anomalies in the electric field, magnetic field, plasma parameters and energetic particles (Pulinets and Boyarchuk 2004; Zhang et al., 2007; Anagnostopoulos and Rigas, 2009). Ionospheric anomalies attract more and more attention nowadays by their shortterm feature, for they always occur one week before earthquakes. A lot of statistical analysis has shown the correlation between the electric field anomalies in the ionosphere and strong earthquakes. An anomalous increase in the intensity of low-frequency (0.1–16 kHz) radiowave emissions was detected by using Intercosmos-19 data (Larkina et al., 1989). Parrot and Mogilevsky (1989) studied the GEOS and AUREOL-3 satellite data and they found that earthquakes caused extremely low frequency electromagnetic emissions in the upper ionosphere. Parrot (1994) analysed the AUREOL-3 satellite data of around 325 earthquakes with Ms > 5. His results showed that during a 24-h window, the maximum amplitude in the electric field occurred in the interval of

ΔLon<10° (Lon~ longitude) from the epicentres regardless of ΔInv. lat (Inv.lat ~ invariant latitude). Molchanov et al. (1993) summarized the 28 earthquakes occurring during 16 November 1989 to 31 December 1989, based on Intercosmos-24 satellite data. They found that emissions with a spectrum maxima were observed at ULF-ELF (f less than 1000 Hz) over the epicentral areas and these emissions were mainly observed at 12–24 h before the main shocks. Serebryakova et al. (1992) found similar EM radiations on satellites COSMOS-1809 and AUREOL- 3 with a frequency below 450 Hz over the seismic region in Armenia before strong earthquakes during 20 January to 17 February 1989. Gousheva et al. (2008) presented their results of anomalies in the quasi-static electric field in the upper ionosphere(h = 800–900 km) observed by the satellite INTECOSMOS-BULGARIA-1300 over seismic regions and found the increase in the vertical component of the electric field based on 250 investigated cases.

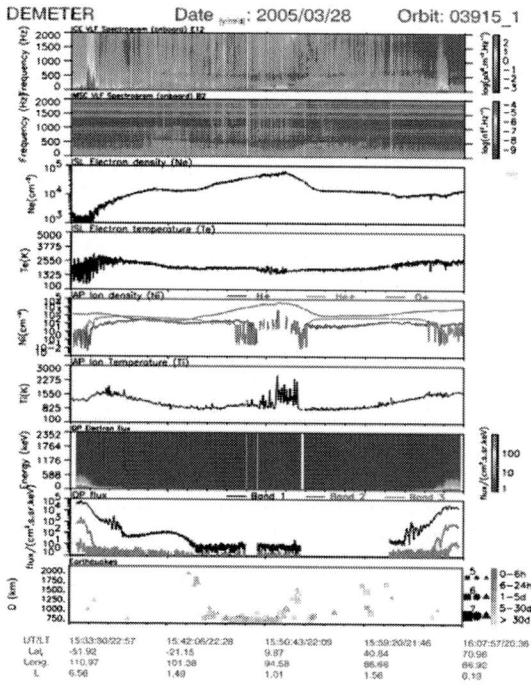

Figure 1: The electromagnetic perturbations before the Sumatra 8.6 Earthquake on 28 March 2005.

DEMETER, a French micro-satellite, launched on 29 June 2004, was the first one in the world to be designed specially for studying the ionospheric variations possibly associated with earthquakes, man made transmitters, volcanoes and lightning, having a solar synchronous circular orbit, declination of 98.23°, and a height of 710 km (which decreased to 660 km in mid-December 2005). A set of instruments were deployed on the satellite, including ICE to detect the electric field from DC to 3.5 MHz; IMSC to measure the magnetic field from a few Hz to 20 kHz; IAP to detect ion density and temperature; ISL, Langmuir probe to measure the electron density and temperature; IDP to detect the energetic electron flux at 72.9 keV–2.34MeV.

As for the study about DEMETER satellite, Parrot et al. (2006) firstly showed examples of ionospheric perturbations in the electron density, electric and magnetic field, and high energy particles before some strong earthquakes. Nˇˊemec et al. (2009) present their results that the power spectra density in the ELF/VLF electric field decreased 0–4 h before the main shocks at the frequency 1.7 kHz during local nighttime. Athanasiou et al. (2011) studied the ULF electromagnetic waves around the Haiti M = 7.0 earthquake, and they exhibited the variations of the Ez-electric field component during a time period of 100 days before and 50 days after it. Their results showed a significant increase in energy Ez for the time interval of 30 days before this earthquake. In some case studies, an interesting phenomenon of electrostatic perturbation was observed in the electric field at a frequency lower than 250 Hz before some strong earthquakes (Zhang et al., 2009; Ouyang et al., 2007), in which the electrostatic perturbations appeared where they rarely occurred over regions at mid- and lower-latitude of 20–35°. In this paper, we are focusing on, in more detail, the electric field data in the nighttime observed by DEMETER around the earthquakes of Ms ≥7.0 in globe since 1 January 2005 to 28 February 2010. The electrostatic perturbations in the ionosphere are picked up and analysed. The coupling mechanism between earthquakes and ionospheric electrostatic perturbations is discussed as well.

DATA COLLECTION AND ANOMALY IDENTIFICATION

Due to incomplete orbits at the end half of 2004, the beginning stage of DEMETER, the satellite data during this period was not included in this paper. A total of 69 strong earthquakes above a magnitude 7.0 since January of 2005 to February 2010 were collected (http://neic.usgs.gov; http://www.csndmc.ac.cn). If the earthquake with a magnitude above 7.0 is dealt with, the scope of the so called "seismic preparation region" could exceed 1000 km from the epicentre expressed in terms of the equation of $\rho = 10^{0.43M}$ based on the statistics of the ground observation (Dobrovosky et al., 1979) within the scope, some ionospheric perturbations could possibly be triggered by the earthquake preparation process. The observing satellite data were chosen during the 9 days with 7 days before and 1 day after these earthquakes. It should be noted that ULF electromagnetic activity was detected much earlier and probably suggests a long period of ULF seismic precursory signals (Hayakawa et al., 1996). Here, we focused on short-term ULF/ELF phenomena after considering the conclusions about ionospheric precursors that occurred mostly within a week before the earthquakes (Pulinets and Boyarchuk, 2004). Only up-orbits recorded during nighttime, under quiet electromagnetic condition, were selected in this paper to avoid the effects of solar activity. Taking into account the spatial correlation, the anomalies could be considered to be related to the earthquakes if they occurred at a latitude scale within ±10° relative to the epicentre latitude, because whatever the direct projection position of the epicentre or signals propagating along the magnetic field line from the focal area may not exceed more than 10° in latitude over the mid-low latitude regions. The pictures in the paper of Gousheva et al. (2008) also verify that ULF electric field anomalies in the ionosphere were distributed over the seismic regions, sometimes extending far from the epicentral latitudes. In order to ensure that at least one orbit of the DEMETER satellite can be found every day over a certain region, the up-orbits were chosen within the distance of 2000 km to the epicentre in longitude.

The orbit (3915-1) in Fig. 1 is such an example of anomalies chosen according to the requirement, just flying over Indonesia 50 min before the Sumatra M =8.6 Earthquake occurring at 16:39:36.52,

28 March 2005, located at 97.11°E– 2.09° N. In Fig. 1, the panels show the parameters (from top to bottom) as follows: the VLF electric field spectrum at 19.5 Hz–2 kHz, the VLF magnetic field spectrum at 19.5 Hz– 2 kHz, Ne(electron density), Te(electron temperature), Ni(ion density of H^+, He^+, O^+), Ti(ion temperature), the energetic electron spectrum between 72.9 keV–2.34MeV, the electron flux at three bands (90–600 keV; 0.6–1MeV; 1–2.34 MeV), the earthquakes occurring less than 2000 km apart from this orbit during ±30 days. As presented in Fig. 1, the disturbances were detected at the equatorial region in most observing parameters, including the electric field at ULF/ELF frequency band that we paid attention to in this paper, electron density, electron temperature, ion density, and so on. Here the perturbations, at ULF/ELF band less than 250Hz of the electric field, are considered as electrostatic perturbations.

Figure 2 exhibits the disturbances extracted from the electric field spectrum along orbit 3915-1. The first top panel presents the median value of PSD (power spectrum density) in the electric field at 19.5–250 Hz, in which the spectra density over the seismic region increased with two orders of magnitude relative to its surrounding normal level, exceeding $10^2 \mu V^2/m^2/Hz$. After repetitive testing, the electrostatic perturbations were selected by an automatic technique developed by Zhang et al. (2010). They were picked following this definition: A0 $>10^{0.7} \mu V^2/m^2/Hz$ where A_0 is the first PSD value at 19.5 Hz, and other PSD values at the following frequency points should fit the exponential relation of $S_E = A_0 \cdot f^{-b}$, where S_E represents the PSD value at a different frequency and f is frequency (Zhang et al., 2010). The lower panel shows the selected anomalous signals (with value 1) and normal points (with value 0). It can be found that the distinguished anomalies are consistent with the perturbations shown in the top panel. In Fig. 2, some strong perturbations were also very clear and significantly above the latitude of 40°, which can be detected almost every day, especially along the orbits of the Eastern Hemisphere. So these perturbations at higher latitudes may not be related to strong earthquakes, but to the auroral electrojets and energetic electrons precipitation into radiation belts over this region, which can be proven by the stronger particle flux at the same latitudes. These signals at high latitudes should be cast off in anomaly identification according to their geographical positions, therefore, earthquakes at high latitudes were not discussed in this paper because it is difficult to distinguish whether or not the perturbations in the electric field were motivated by earthquakes under this strong noise background.

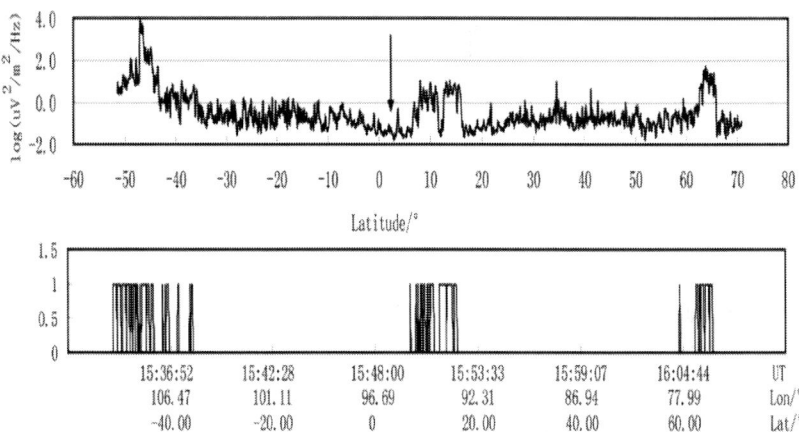

Figure 2: Ionospheric perturbations in electric field along the orbit 3915-1.

Based on the definition of anomalies and electrostatic perturbations, among 69 earthquakes within the latitude of 40°, the electrostatic perturbations were observed before the 32 earthquakes along the up-orbits in a distance from the epicentres less than 2000 km during the 9 days, while there is one earthquake only showing post-seismic anomaly without precursors. All the details about these 32 earthquakes and anomalies around them are listed in Table 1, including the date, time, magnitude(Ms), longitude(Lon), latitude(lat), depth of the earthquake and also the time differences between anomalies and earthquakes, latitude and longitude scale of anomalies, median value of PSD in the electric field at 19.5– 250 Hz, and daily $\sum$ Kp index. Some of the $\sum$ Kp values are followed by the letter D that represents that day being in a disturbed state. Actually, the Kp in most days was not marked by D in Table 1, which means most anomalies appeared in quiet electromagnetic condition. It also demonstrated, from another aspect, that the perturbations could be well correlated with earthquakes, instead of solar activity or other space impact-factors due to low Kp values during those days.

It can be seen from Table 1 that the anomalies in ULF/ELF electric field appeared many times before some earthquakes, such as the Sumatra 8.6 Earthquake on 28 March 2005; Haiti 7.0 Earthquake on 12 January 2010 and so on, which reflect that ULF/ELF perturbations continued a long time over the seismic region during the earthquake preparation process. Moreover, there are 21 earthquakes among 32 events with the ionospheric perturbations occurring in 3 days prior to them, showing the short-term temporal feature of ULF/ELF EM perturbations in ionosphere.

ANALYSIS ON ELECTROSTATIC PERTURBATIONS AND DISCUSSION

Spatial Distribution of Earthquakes

Figure 3 shows the global epicentral distribution of the selected 32 earthquakes (Fig. 3a), and the projection of their depths versus latitudes (Fig. 3b). It can be seen that: (1) these earthquakes are mostly located at the plate boundaries; (2) there are 21 earthquakes located in the latitude scale of $\pm 20°$, with electrostatic perturbations being unusual; (3) among them 8 earthquakes are deeper than 100 km, occupying 25 %, and even most earthquakes are located in the ocean, not on land, as shown in Table 1. Figure 3a shows that these 32 earthquakes are mainly along the Circum-Pacific Seismic belt, which may indicate that interaction between giant plates would more easily produce intensive anomalies along the major faults and then induce electrostatic perturbations in the ionosphere.

It is well known that when EM waves propagate in water from the ocean bottom, they will attenuate largely, that is to say, the EM wave has difficulty when penetrating seawater into the atmosphere and ionosphere directly. Why so many perturbations were detected before oceanic earthquakes? A speculation might explain it as follows: radon or other chemical materials emitted from the oceanic faults would change the ionization in seawater and the composition of water ions in it. This change would lead to the variation of the atmospheric vertical electric field over the seismic region. The variation of atmospheric electric field as a part of the current system between ionosphere and lithosphere would cause the change of the current system in the ionosphere, which might lead to the disturbances in kinds of parameters including the ULF/ELF electric field.

Table 1: Summary of anomalous information in ULF/ELF electric field related to strong earthquakes

Date y-m-d	Time h-m-s	Ms	Lon /°E	Lat /°N	Depth /km	Land or Ocean	Δt T-Te	Latitude scale of anomalies	Longitude scale of anomalies	Median value -lg(μV²/m²/Hz)	Kp
2010-02-27	06-34-16.4	8.8	-72.7	-35.8	33	L	-3 day	39°~42°S	298.5°~299.5°E	1	6
							-2 day	39°~41°S	291°~292°E	0.5	6
2010-1-12	21-53-09.85	7.0	-72.53	18.46	10	L	-5 day	23°~26°N	302.5°~303°E	0~1	1
							-1 day	5°~20°N	273°~275°E	0~1	13
							+1 day	10°~21°N	282°~286°E	0~0.5	14(D)
2010-1-3	22-36-28.15	7.2	157.3	-8.9	25	O	-12 h	4°~12°S	168°~169.5°E	0.5	7
2009-10-7	22-18-51.24	7.8	166.38	-12.52	35	O	-1 day	1°~9°S	176°~177.5°E	0.5~1	3
2009-9-29	17-48-10.99	8.1	-172.1	-15.49	18	O	-5 day	6°~16°S	194°~195°E	1~2	2
2009-9-2	07-55-01.05	7	107.3	-7.78	46	O	-2 day	10°~16°S	106°~107°E	0~0.5	11
2009-8-10	19-55-35.61	7.5	92.89	14.1	4	O	-1 day	12°~14°N	95°~96°E	0~1	12
2009-8-9	10-55-55.61	7.1	137.94	33.17	297	O	-4 day	35°~43°N	117°~129°E	1	11
							+1 h	32°~37°N	138°~140°E	0.5	12
2009-3-19	18-17-40.91	7.6	-174.66	-23.05	34	O	-6 day	17°~24°S	169°~171°E	0.5~1	25(D)
							-1 day	18°~20°S	180°~181°E	0.5	3
2009-2-18	21-53-45.16	7.0	-176.33	-27.42	25	O	-5 day	10°~21°S	182°~185°E	0.5~1.2	4
							-4 day	12°~20°S	175°~177°E	1	24(D)
							-2 day	16°~28°S	185°~188°E	1	7
							-18.5 h	18°~26°S	195°~197°E	0.5	6
2009-2-11	17-34-50.49	7.2	126.39	3.89	20	O	-3 day	6°S~2°N	119°~121°E	0.5	1
							+1 day	0°~5°N	137°~138°E	0.5~1.0	3
2008-4-9	12-46-12.72	7.3	188.89	-20.07	33	O	-5 day	12°~20°S	159°~161°E	0.4	15
							-2 h	16°~21°S	170°~171.5°E	0.5	22
2007-12-9	07-28-20.82	7.8	-177.51	-26	152	O	-7 day	18°~19°S	179°~180°E	1.5	3
							-5 day	28°~35°S	191°~193°E	1~1.5	2
							-3 day	20°~30°S	175°~200°E	0.5~1.2	5
							-2 day	24°~32°S	190°~200°E	0.5~1.2	2
2007-11-29	19-00-20.42	7.4	-61.27	14.94	156	O	-5 day	4°~6°N	309°~309.5°E	0~1	22(D)
2007-11-14	15-40-50.53	7.7	-69.80	-22.25	40	L	-6 day	24°~37°S	291°~295°E	0	7
							-3 day	25°~40°S	294°~297°E	0	4
							-1 day	10°~23°S	299°~301°E	-1~0	19
2007-9-12	11-10-26.83	8.5	101.37	-4.44	34	L	-5 day	4°~10°S	100°~101°E	0.7~1.5	18
							-5.7 h	8°~10°S	111°~112°E	0.3	4
2007-9-2	01-05-8.15	7.2	165.76	-11.61	35	O	-5 day	6°~16°S	177°~179°E	0.5~1	18
							-4 day	5°~20°S	169°~173°E	0.5~1	9
2007-08-15	23-40-57.89	8.0	-76.6	-13.39	39	L	-6 day	8°~12°S	298.5°~299.3°E	0.5	5
2007-8-8	17-05-04.92	7.5	107.42	-5.86	280	O	-4 day	1°S~5°N	113°~115°E	0.5~1.2	2
							-2 day	4°~9°S	100°~101°E	0.5~1	18
							-2 h	4°S~5°N	107°~110°E	0.5~1	11
2007-8-1	17-08-51.4	7.2	167.68	-15.6	120	O	-6 day	10°~20°S	161°~163°E	0.5~1.5	13
2007-4-1	20-39-58.71	7.2	157.04	-8.47	24	O	-4 day	6°~10°S	149°~150°E	0.5~1.5	12
2007-3-25	00-40-1.61	7.1	160.36	-20.62	34	O	-7 day	4°~15°S	151°~154°E	0.5	6
							-5 day	4°~18°S	161°~164°E	0.5~1.5	1
2006-7-17	08-19-30.5	7.3	107.4	-9.4	20	O	-7 day	18°~20°S	116.5°~117°E	1	17
							-6 day	14°~20°S	108°~118°E	1	13
							-5 day	14°S~6°N	96°~101°E	0~1	19(D)
							4 day	20°S~20°N	90°~120°E	0~1	10
							-3 day	20°S~20°N	100°~115°E	1	19(D)
							+7 h	14°~20°S	111°~113°E	1~1.5	6
2006-5-16	15-28-24.6	7.2	97.2	6.1	12	O	-1 h	2°S~2°N	117°~118°E	0.5~1	6
							+0.5 h	8°~10°S	94°~96°E	0.4~1	6
2006-5-16	10-39-29.4	7.4	-179.31	-31.81	152	O	-1 h	33°~35°S	198°~200°E	0.5~1	6
2006-2-22	22-19-9.6	7.5	33.2	-21.3	11	L	-3 day	9°~12°S	42°~43°E	0.5~1	14
2006-1-27	16-58-50.6	7.6	128.1	-5.4	397	O	-2.5 h	2.8°S~15°N	115°~126°E	0~1	20
2005-9-26	01-55-37.67	7.5	-76.4	-5.68	115	L	-4 day	14°~25°S	279°~279°E	1	13
							-3 day	10°~25°S	200°~266°E	1	12
2005-9-9	07-26-43.73	7.6	153.47	-4.54	90	O	-5 day	6°~10°S	147°~148°E	0.5~1	29
							-4 day	4.5°~6°S	160.5°~161°E	0.4~0.6	20
							-3 day	10°S~19°N	144°~154°E	0.5~1	15
							-1 day	6°~10°S	153.5°~154.5°E	0.5~1	13
							+3.5 h	4°~13°S	167°~168°E	0.4~1	22
2005-08-16	02-46-28.4	7.2	142.04	38.28	36	L	+1 day	37°~41°N	157°~159°E	0~1	22
2005-07-24	15-42-06.21	7.2	92.19	7.92	16	O	-2 day	14°~15°N	80°~81°E	0.5~1	20

2005-3-28	16-09-36.53	8.6	97.11	2.09	30	L	−6 day	1° S~4°N	109°~110° E	1~1.5	5
							−5 day	6° S~2°N	101°~103° E	0.5~1	7
							−3 day	8°~11°N	102°~103° E	0.3	27(D)
							−20 min	3.5°~14.8° N	93°~95° E	0.5~1	9

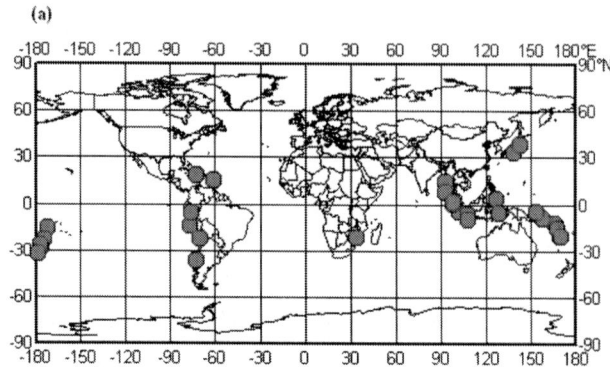

(a)

(b)

Figure 3: Spatial (a) and depth (b) distribution of earthquakes with electro-static perturbations (the circle represents the earthquake).

Extended Spatial Characteristic of ULF/ELF Electric field Perturbations

In order to study the relationship between ULF/ELF electrostatic perturbations and earthquakes, the global distribution of the perturbations before the Sumatra M =8.6 Earthquake on 28 March 2005 were selected as an example and the time differences were still limited to 7 days before the earthquakes as mentioned above. As shown in Table 1, the perturbations appeared on 22, 23 25 and 28 March in the orbits within the range of 2000 km from the epicentre. Here the question is whether the anomalies in the ionosphere only occur at the orbits nearest to the epicentre, or also at other orbits. Take the Sumatra Earthquake on 28 March 2005 as an example, the black segments marked in Fig. 4 give the global distribution of EM perturbations during those days except the ones on 25 March which might be influenced by magnetic storms on that day. To allow a convenient comparison and visualized figures, the electric field signals with PSD larger than 5 $\mu V^2 \cdot m^{-2} \cdot Hz^{-1}$ and fitting exponential delay laws at the low frequency band of 19.5–250 Hz (Zhang et al., 2010) were assigned values of 1 (black circles) and taken as a time segment of perturbations, or else they would be assigned values of 0 (gray circles) and taken as a time segment without disturbances. All the perturbations at the ultra-low frequency band are picked up along the orbits and plotted in Fig. 4. It can be seen that, at the latitude higher than 40°, there existed lots of these ULF/ELF perturbations, reasonably due to the effects of the polar ring current, energetic electron precipitation and other factors, which can also be seen very clearly in Figs. 1 and 2. Besides those signals at high latitudes, one could, however, find many perturbations around the epicentre of this M =8.6 earthquake (black triangle) at lower latitudes in the range of ±20°, and they extended from 0° to 150° E on 22 and 23 March. While, the perturbations on 28 March only occurred at the orbit closest to the epicentre, which shows a different feature with those on 22 and 23.

Similar EM perturbations, with a large scale, were also found before some other earthquakes (Zhang et al., 2010; Ruzhin et al., 1998). It seems to the authors that the following hypothesis of taking the stress changes during the earthquake preparation process as a source corresponding to the perturbations in the ionosphere, would be helpful in understanding the feature. During the seismic preparation stage, there would be a region much larger than the epicentral area, being under a state of stress accumulation, and the electromagnetic signals might be frequently produced and continue for a long time. When they propagate to the ionosphere, together with the ion and electron drift in the ionosphere, they can be detected by many orbits. As soon as the earthquake process enters into its impending stage, the stress would be concentrated just in the epicentral area while the anomalous region would shrink correspondingly.

Discussion

The possible coupling mechanisms among lithosphereatmosphere-ionosphere (LAI) have been suggested in many publications (Pulinets et al., 2000; Pulinets and Boyarchuk, 2004; Molchanov et al., 2004; Rycroft, 2006; Namgaladze et al., 2009; Pulinets and Ouzounov, 2011). Some of them will be discussed on the ionospheric perturbations in the ULF/ELF electric field associated with strong earthquakes.

One possible mechanism is the direct penetration of ULF/ELF EM waves from the epicentres into the ionosphere. There are many results to illustrate the existence of ULF/ELF/VLF electromagnetic (EM) emissions prior to strong earthquakes in ground-based observations (Hayakawa, 2004). When energy is accumulated underground to some extent, microstructures will increase and electromagnetic emissions will be produced simultaneously. Another source has also been proposed that positively charged holes are easy to constitute into minerals especially semi-conductor minerals when they are heated. Some laboratory experiments support this possibility (Freund, 2000; Shen et al., 2009). Based on numerical computations, ULF/ELFelectromagnetic emissions lower than 20 Hz can transverse directly into the upper ionosphere by penetrating or nonpenetrating solutions with a decrease of an order of magnitude (Bortnik and Bleier, 2004). The transverse electric (TE) component of ULF noises is transformed into Alfven waves at the atmosphere-ionosphere boundary, and nonlinear interactions of ULF Alfven waves with energetic particles in the magnetosphere may result in the occurrence of ELF/VLF emissions in the upper ionosphere (Larkina et al., 1989; Molchanov et al., 1993).

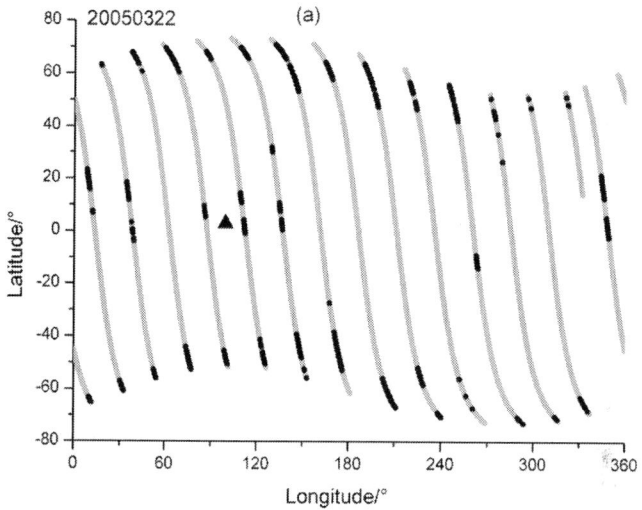

(a)

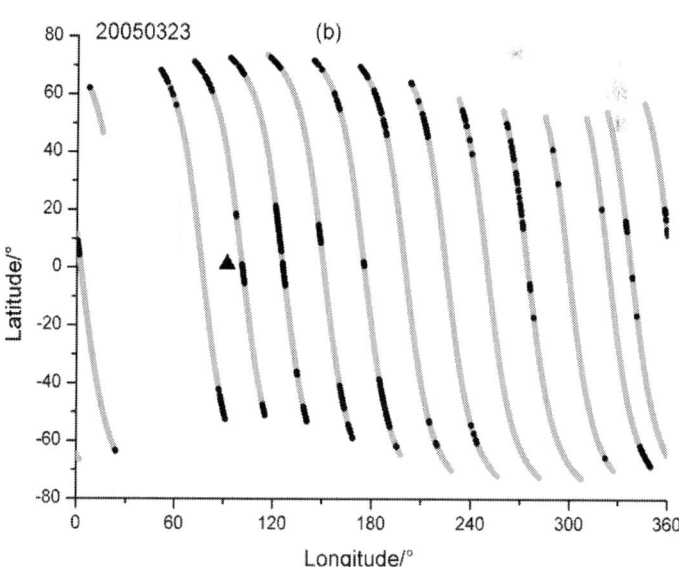

(b)

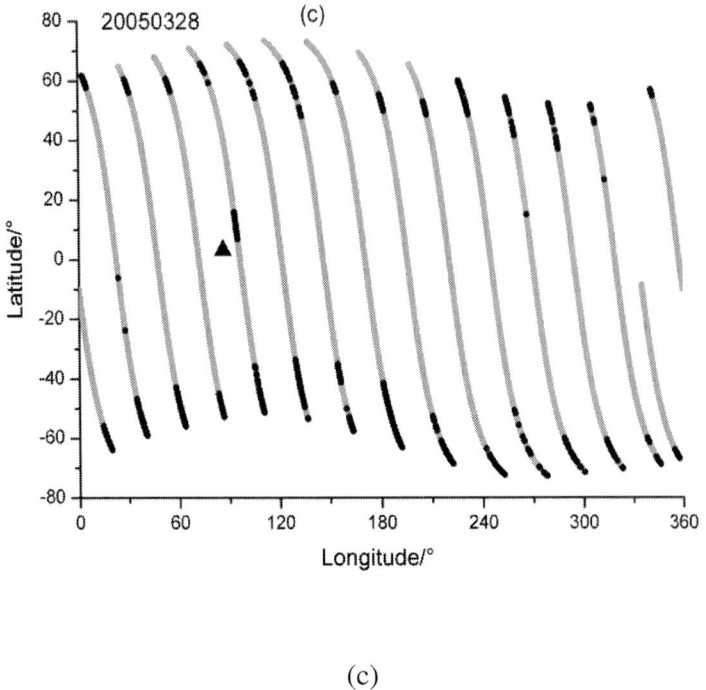

(c)

Figure 4: The spatial distribution of electrostatic perturbations (dark points) on 22, 23 and 28 March before a 8.6 earthquake at Sumatra Indonesia on 28 March 2005 (the triangle in the figure is the epicentre).

Another one is that the enhanced Equatorial Ionospheric Anomaly (EIA) may also induce electrostatic (ES) turbulences (Namgaladze et al., 2009). The heating of sunlight and tidal effects will lead to the upward movement of plasma in the lower ionosphere, penetrating the geomagnetic power lines and then constructing an electric current in E layer. This electric current acts with horizontalmagnetic power lines and causes the increase in electron density in the ionosphere at ±20° geomagnetic latitudes around the magnetic equator. In our paper, at the observing altitude of DEMETER with 660– 710 km, the ionospheric crest is always shown as one peak near the magnetic the equatorial area (see the first panel of electron density in Fig. 1), but as shown in Fig. 4 and combined with the results of Pulinets et al. (2006), the perturbations occurred at two sides of the crest of Ne on 22 and 23 March before the Sumatra Earthquake. Over an epicentral area near the equator, the vertical electric field might be changed due to the accumulation of radon or aerosols in the near-earth atmosphere at the seismic preparation area (Liperovsky et al., 2005), and then the east directed polarization electric field would be generated due to the different drift velocities of electrons and ions. The east directed electric field would stimulate the equatorial anomaly amplification while plasma bubble would be formed in the bottom-side ionosphere and float up to the DEMETER altitude (Pulinets et al., 2006). The double peak structure can persist until the late evening hours, just the time of the DEMETER up-orbits in local nighttime, which has been recently proven in the paper of Vyas and Andamandan (2011). So this mechanism may be used to explain some anomalous phenomena in ULF/ELF electric field near the equatorial area.

CONCLUSIONS

Among 69 earthquakes, DEMETER satellite observed ionospheric perturbations in the ULF/ELF electric field during local nighttime before the 32 earthquakes, which demonstrates that these ionospheric disturbances were not casual phenomena, but may be associated with earthquakes. All the characteristics of these anomalies were summed up as follows:

- Before 46%strong earthquakes in 69 studied cases with a magnitude above Ms = 7.0, the electrostatic perturbations were obtained at 19.5–250 Hz in a distance of2000 km and latitudinal difference of 10° in the ionosphere. The anomalies occurred mostly within 3 days before the 21 earthquakes. Only 7 earthquakes among them showed anomalies a few hours before the earthquakes (Table 1). These 32 earthquakes are mainly located at the boundary of plates. But the depth of earthquakes does not show significant influence on the forming of ES turbulences in the ionosphere.

- Extended study in this paper proved that before the 8 earthquakes, the perturbations in the ionosphere could be observed in a very large scale in longitude, but when the observing time was closer to the earthquake occurrence, the anomalous area shrank, and perturbations always only occurred along the closest orbits apart from the epicentres, which may be related to the different stress developing stages in the earthquake preparation process.

- There are 54% of earthquakes with no obvious ES perturbations detected. The main reasons that are taken into account: the first, some of these cases are located at high latitudes so the ES perturbations can not be easily distinguished with those long existing ES turbulences at this region; the second, the flying time of a single satellite is limited when it crosses a certain place, only once a day like the DEMETER satellite, so it can not be ensured that the anomaly at the seismic region can continue for a very long time in order to meet the satellite; the third, there was no anomaly at all at the seismic region, or the anomaly is not intense enough to result in ionospheric perturbations.

- The LAI coupling process is complex and there are only some qualitative interpretations at present. Based on the results in this paper and combined with other researches, the direct ULF/ELF EM propagation, the coupling mechanism between the enhanced vertical electric field in the atmosphere and EIA amplification are suggested to be important factors to explain the ionospheric electrostatic perturbations in the ULF/ELF electric field. In order to understand and verify the mechanism between the ionospheric perturbations with strong earthquakes, it is necessary to strengthen the observation of multi-parameters on the ground, in the atmosphere and ionosphere synchronously.

ACKNOWLEDGEMENTS

This paper is funded by the International Cooperation Project (2009DFA21480) and the Key Earthquake Science Research Fund (201008007). We are grateful to the DEMETER Data Centre for the provision of the satellite data.

REFERENCES

1.   Anagnostopoulos, G. and Rigas, V.: Variations of energetic radiation belt electron precipitation observed by DEMETER before strong earthquakes, Geophys. Res. Abstr., EGU2009-10700, 11, 2009.

2. Athanasiou, M. A., Anagnostopoulos, G. C., Iliopoulos, A. C., Pavlos, G. P., and David, C. N.: Enhanced ULF radiation observed by DEMETER two months around the strong 2010 Haiti earthquake, Nat. Hazards Earth Syst. Sci., 11, 1091–1098, doi:10.5194/nhess-11-1091-2011, 2011.

3. Bortnik, J. and Bleier, T.: Full wave calculation of the source characteristics of seismogenic electromagnetic signals as observed at LEO satellite altitudes, Eos Trans. AGU, 85(47), Fall Meet. Suppl., Abstract T51B-0453, 2004.

4. Dobrovosky, I. R., Zubkov, S. I., and Myachkin, V. I.: Estimation of the size of earthquake preparation zones, PAGEOPH, 117, 1025– 1044, 1979.

5. Freund, E.: Time-resolved study of charge generation and propagation in igneous rocks, J. Geophys. Res., 105, 11001– 11019, 2000.

6. Gousheva, M., Danov, D., Hristov, P., and Matova, M.: Quasi-static electric fields phenomena in the ionosphere associated with preand post earthquake effects, Nat. Hazards Earth Syst. Sci., 8, 101–107, doi:10.5194/nhess-8-101-2008, 2008.

7. Hayakawa, M.: Electromagnetic phenomena associated with earthquakes: a frontier in terrestrial electromagnetic noise environment, Recent Res. Dev. Geophys., 6, 81–112, 2004.

8. Hayakawa, M., Molchanov, O. A., Ondoh, T., and Kawai, E.: The precursor signature effect of the Kobe earthquake in VLF subionospheric signal, J. Comm. Res. Lab., Tokyo, 43, 160, 1996.

9. Larkina, V. I., Migulin, V. V., Molchanov, O. A., Kharkov, I. P., Inchin, A. S., and Schvetcova, V. B.: Some statistical results on very low frequency radiowave emissions in the upper ionosphere over earthquake zones, Phys. Earth Planet. In., 57, 1–2, 100–109, 1989.

10. Liperovsky, V. A., Meister, C.-V., Liperovskaya, E. V., Davidov, V. F., and Bogdanov, V. V.: On the possible influence of radon and aerosol injection on the atmosphere and ionosphere before earthquakes, Nat. Hazards Earth Syst. Sci., 5, 783–789, doi:10.5194/nhess-5-783-2005, 2005.

11. Molchanov, O. A., Mazhaeva, O. A., Goliavin, A. N., and Hayakawa, M.: Observation by the Intercosmos-24 satellite of

ELF-VLF electromagnetic emissions associated with earthquakes, Ann. Geophys., 11, 431–440, 1993, http://www.ann-geophys. net/11/431/1993/.

12. Molchanov, O., Fedorov, E., Schekotov, A., Gordeev, E., Chebrov, V., Surkov, V., Rozhnoi, A., Andreevsky, S., Iudin, D., Yunga, S., Lutikov, A., Hayakawa, M., and Biagi, P. F.: Lithosphereatmosphere-ionosphere coupling as governing mechanism for preseismic short-term events in atmosphere and ionosphere, Nat. Hazards Earth Syst. Sci., 4, 757–767, doi:10.5194/nhess-4-757- 2004, 2004.

13. Namgaladze, A. A., Klimenko, M. V., Klimenko, V. V., and Zakharenkova, I. E.: Physical mechanism and mathematical modeling of earthquake ionospheric precursors registered in total electron content, Geomagn. Aeronomy+, 49, 2, 252–262, 2009.

14. Nˇemec, F., Santol´ık, O., and Parrot, M.: Decrease of intensity of ELF/VLF waves observed in the upper ionosphere close toearthquakes: A statistical study, J. Geophys. Res., 114, A04303, doi:10.1029/2008JA013972, 2009.

15. Ouyang, X. Y., Zhang, X. M., Shen, X. H., Liu, J., Qian, J. D., Cai, J. A., Zhao, S. F.: Ionospheric Ne disturbances before 2007 Pu'er Yunnan China earthquake, Acta Seismologica Sinica, 21(4), 425–437, 2008.

16. Parrot, M.: Statistical study of ELF/VLF emissions recorded by a low-altitude satellite during seismic events, J. Geophys. Res., 99(A12), 23339–23347, 1994.

17. Parrot, M., Berthelier, J. J., Lebreton, J. P., Sauvaud, J. A., Santol´ık, O., and Blecki, J.: Examples of unusual ionospheric observations made by the DEMETER satellite over seismic regions, Phys. Chem. Earth, 31, 486–495, doi:10.1016/j.pce.2006.02.011, 2006.

18. Parrot, M. and Mogilevsky, M. M.: VLF emissions associated with earthquakes and observed in the ionosphere and the magnetosphere, Phys. Earth Planet. In., 57, 1–2, 86–99, 1989.

19. Pulinets, S. A., Boyarchuk, K. A., Hegai, V. V., Kim, V. P., and Lomonosov, A. M.: Quasielectrostatic model of atmospherethermosphere- ionosphere coupling, Adv. Space Res., 26, 1209– 1218, 2000.

20. Pulinets, S. A. and Boyarchuk, K. A.: Ionospheric Precursors of Earthquakes, Springer, Berlin, Heidelberg, New York, 1–287, 2004.

21. Pulinets, S. and Ouzounov, D.: Lithosphere-Atmosphere-Ionosphere Coupling(LAIC) model-An unified concept for earthquake precursors validation, J. Asian Sci., 41, 371–382, 2011.

22. Pulinets, S., Ouzounov, D., and Parrot, M.: Conjugated nearequatorial effects registered by DEMETER satellite before Sumatra earthquake M = 8.7 of March 28, 2005, International Symposium of DEMETER, Toulous-France, 14–16 June 2006.

23. Ruzhin, Y. Y., Larkina, V, A., and Depueva, A. K.: Earthquake precursors in magnetically conjugated ionosphere regions, Adv. Space Res., 21, 525–528, 1998.

24. Rycroft, M. J.: Electrical processes coupling the atmosphere and ionosphere: An overview, J. Atmos. Solar-Terr. Phys., 68, 445–456, 2006.

25. Serebryakova, O. N., Bilichenko, S. V., Chmyrev, V. M., Parrot, M., Rauch, J. L., Lefeuvre, F., and Pokhotelov, O. A.: Electromagnetic ELF radiation from earthquake regions as observed by low-altitude satellites, Geophys. Res. Lett., 19, 91–94, 1992.

26. Shen, J. F., Shen, X. H. and Liu, Q.: Thermoelectricity property of natural semi-conductor minerals and its application in earthquake prediction, Bulletin of Mineralogy, Petrology and Geochemistry, 28, 301–307, 2009,

27. Vyas, B. M. and Andamandan B.: Nighttime VHF ionospheric scintillations characteristic near the crest of Appleton anomaly station Udanpur (24.6_ N, 73.7_ E), Indian J. Radio Space, 40, 191–202, 2011.

28. Zhang, X., Battiston, R., Shen, X., Zerenzhima, Ouyang, X., Qian, J., Liu, J., Huang, J., and Miao, Y.: Automatic collecting technique of low frequency electromagnetic signals and its application in earthquake study, in: Knowledge Science, Eng. Manag., edited by: Bi, Y. and Williams, M. A., KSEM2010, LANI 6291, Springer, Berlin, Heidelberg, 366–377, 2010.

29. Zhang, X., Qian, J., Ouyang, X., Shen, X. H., Cai, J. A., and Zhao, S. F.: Ionospheric electromagnetic perturbations observed on

DEMETER satellite before Chile M7.9 earthquake, Earthquake Sci., 22, 251–255, 2009.

30. Zhang, X., Zhao, G. Z., Chen, X. B., and Ma, W.: Seismoelectromagnetic observation abroad, Prog. Geophys., 22, 3, 687– 694, 2007

Chapter 3

Possible Electromagnetic Effects on Abnormal Animal Behavior before an Earthquake

Masashi Hayakawa [1, 2, 3]

[1]Hayakawa Institute of Seismo Electromagnetics Co. Ltd., University of Electro-Communications (UEC) Incubation Center, 1-5-1 Chofugaoka, Chofu, Tokyo 182-8585, Japan

[2]Advanced Wireless Communications Research Center, UEC, Chofu, Tokyo 182-8585, Japan

[3]Earthquake Analysis Laboratory, Kita-aoyama 2-12-42-R305, Minato-ku, Tokyo 107-0061, Japan

ABSTRACT

The former statistical properties summarized by Rikitake (1998) on unusual animal behavior before an earthquake (EQ) have first been presented by using two parameters (epicentral distance (D) of an anomaly and its precursor (or lead) time (T)). Three plots are utilized to characterize the unusual animal behavior; (i) EQ magnitude (M) *versus* D, (ii) log T *versus* M, and (iii) occurrence histogram of log T. These plots are compared with the corresponding plots for different seismo-electromagnetic effects (radio emissions in different frequency ranges, seismo-atmospheric and -ionospheric perturbations) extensively obtained during the last 15–20 years. From the results of comparisons in terms of three plots, it is likely that lower frequency (ULF (ultra-low-frequency, f ≤ 1 Hz) and ELF (extremely-low-frequency, f ≤ a few hundreds Hz)) electromagnetic emissions exhibit a very similar temporal evolution with that of abnormal animal behavior. It is also suggested that a quantity of field intensity multiplied by the persistent time (or duration) of noise would play the primary role in abnormal animal behavior before an EQ.

INTRODUCTION

It is widely reported that land animals, birds, fish *etc.* often respond to earthquakes (EQs). A considerable number of books and papers have been published on this unusual biological behavior prior to EQs (e.g., [1,2,3,4]). Such abnormal animal behavior include: (i) disappearance of rats from a house; (ii) birds crying, *etc.* In addition to these publications, we can add recent works on this topic from the last 10 years or so [5,6,7,8,9,10,11,12]. The papers [7,8] have focused on the abnormal animal behavior for two disastrous EQs (Kobe and Wenchun EQs). These abnormal animal responses are generally called "macroscopic" anomalies of EQs, which are mainly based on anecdotal and retrospective records of animal behavior. Their studies enable us to deduce some physical insight into why and how animals react precursorily to seismic events. In his well-documented book, Rikitake (1998) [4] concluded that one of the most probable mechanisms of biological anomalies seems to be electromagnetic effects. On the other hand, electromagnetic phenomena associated with EQs (sometimes called seismo-electromagnetics) have been extensively investigated during the last 15–20 years for the sake of short-term EQ prediction, and the long-term observations have enabled us to yield some statistical results on different electromagnetic effects (e.g., [13,14]).

In this paper we first present some statistical conclusions by Rikitake (1998) [4] on macroscopic phenomena and then present recent statistical summaries of seismo-electromagnetic effects. Then we try to compare them in order to obtain some indication on the possible electromagnetic effects on biological systems before EQs. Finally we suggest that lower frequency (especially ULF (ultra-low-frequency, f ≤ 1 Hz) and ELF (extremely low frequency, f ≤ 1 kHz) seismogenic emissions, often recorded before an EQ, would be the most promising candidate to explain such unusual biological behavior.

UNUSUAL ANIMAL BEHAVIOR BEFORE AN EQ

In his book on "The science of macro-anomaly precursory to an earthquake", Rikitake (1998) [4] summarized the behavior of animals (small and large), birds *etc.* on the basis of his extensive retrospective analyses for six large EQs including: (i) Ansei-Tokai EQ (M = 8.4, 1854); (ii) Nobi EQ (M = 8.0, 1891); (iii) Kanto EQ (M = 7.9, 1923); (iv) Toh-nankai EQ (M = 7.9, 1944); (v) Izu-oshima off-sea EQ (M = 7.0, 1978); and (vi) Miyagi-oki EQ (M = 7.4, 1978); and also publications by Kayano (1983, 1984) [15,16] on two EQs (Ibaraki-ken Nanbu EQ (M = 6.0, 1978) and Nagano-ken seibu EQ (M = 6.8, 1984)). His summary is presented in terms of two parameters: (1) distance of anomaly from the epicenter (D); and (2) precursory (or lead) time (T).

The statistical results of the macroscopic phenomena are presented in the following three common ways: (i) the relationship of the anomaly between the EQ magnitude (M) and the epicentral distance D; (ii) log T (in units of days) *versus* M; and (iii) the distribution of log T. Since the magnitude M is essentially a logarithmic scale, Figure 1 illustrates the log-log relation of abnormal animal behavior (among different macroscopic phenomena) between M and D [4]. Animals mean here dogs, cats and so on. This figure indicates a tendency that, for larger M values, the precursory anomalous of animal behavior is observed farther from the epicenter of a future EQ. The straight line in the figure—which traces the averaged relation between M-log D on the basis of various types of macroscopic effects (including animals, birds, fish, *etc.*) [4]—is expressed by

M = 1.86 + 2.6 log D   (1)

Figure 2 summarizes the precursor time (T) *versus* M relation. The value of T is distributed over a range from a few minutes to hundreds of days for any specific M. This suggests that there is no clear relationship between T and M. However, the occurrence histogram of log T (in units of days) in Figure 3 indicates that the distribution of T is concentrated in a range of T = 1–10 days.

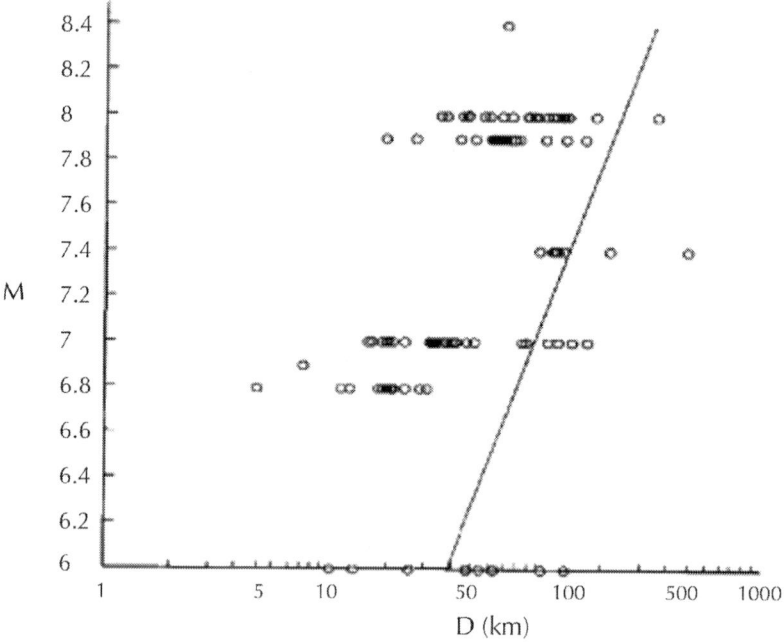

Figure 1: Dependence of unusual animal behavior on the earthquake (EQ) magnitude (M) and the epicentral distance (D). The straight log-log line is the averaged relation between M and D. Adapted from Rikitake (1998) [4] with permission of the publisher.

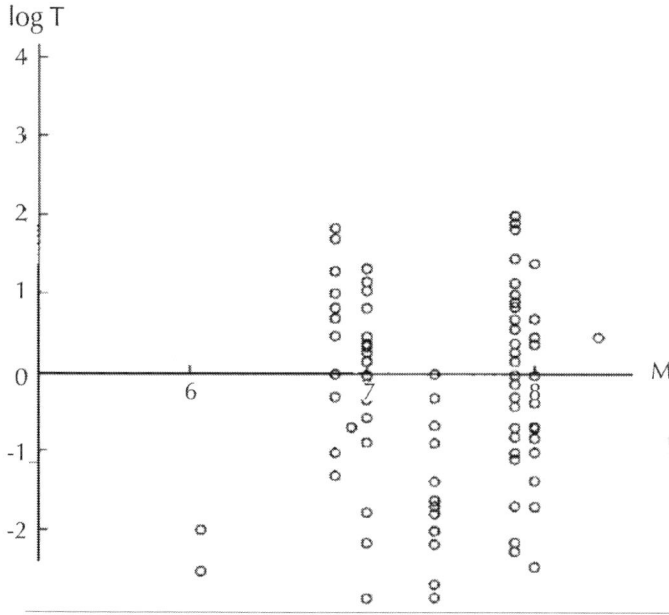

Figure 2: The relationship between M and log T (precursory time in units of days) for unusual animal behavior. Adapted from Rikitake (1998) [4] with permission of the publisher.

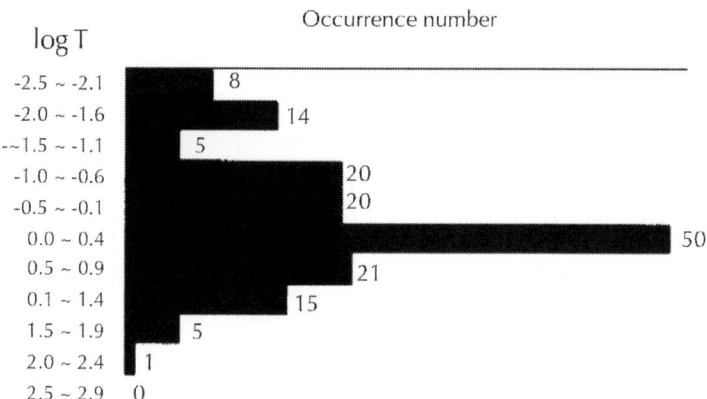

Figure 3: Occurrence histogram of time T (in units of days) of reported unusual animal behavior. Adapted from Rikitake (1998) [4] with permission of the publisher.

Rikitake [4] added another informative statement. In response to the question whether there are any differences in the unusual behaviors between large and small animals, he reports that smaller animals seem to react earlier than larger animals. With respect to birds and fish, Rikitake [4] concluded that nearly all distribution of their unusual behavior is similar to those shown in Figure 1, Figure 2 and Figure 3. Similar results have been reported for snakes, earthworms, insects, *etc.*

As is seen from Figure 1 and Figure 2, Rikitake [4] did his analysis for relatively large EQs with M ≥ 7.0 because of the specific nature of the macroscopic pre-EQ anomalies. It is highly likely that the macroscopic data for lower M EQs would include more inaccurate information (or noise) to the macroscopic data for lower M values. At the same time, we need to emphasize here that most data on electromagnetic phenomena presented below were collected for M values smaller than 7 except for a few data on ULF emission, which tends to occur only for large EQs with M larger than 6–7.

POSSIBLE SENSORY MECHANISM OF ANIMALS

First of all, it seems highly plausible that animals behave unusually prior to EQs, Therefore, as the next step we have to ask what kind of stimuli are likely to lead to unusual animal behavior. Based on the extensive previous studies by Evernden (1976) [1], Buskirk *et al.* (1981) [2], Tributsch (1982) [3], and Rikitake [4], the following is a list of possible candidates of EQ precursory phenomena acting as stimuli:

- Change in atmospheric pressure
- Change in gravity
- Ground deformation (ground uplift and tilt change)
- Acoustic signals and vibrations due to the generation of microcracks
- Electromagnetic effects
- Ground water level change
- Emanation of gases and chemical substances

Based on the available evidence, Rikitake [4] concluded that the most probable candidate for abnormal animal behavior might be (5) electromagnetic effects, though some others, for example (4) and (7), cannot be ruled out. Recently Grant *et al.* [12] have discussed the effect of item (7).

ELECTROMAGNETIC EFFECTS AND THEIR STATISTICAL PROPERTIES

The history of the study of seismogenic electromagnetic effects is rather short, on the order of a few decades, but there has been much progress with respect to short-term EQ prediction, especially since the 1995 Kobe EQ (e.g., see books [13,14,17,18,19,20] or review papers [21,22]). The observation of seismogenic effects can be customarily classified into two categories: (1) direct effect of electromagnetic emissions from within the lithosphere; and (2) indirect effects in the atmosphere or ionosphere. The summaries of different phenomena belonging to both categories will be discussed one by one in relation to the previous three common relationships of M *versus* D, T *versus* M, and occurrence histogram of T.

Seismogenic Radio Emissions

DC Geoelectric Field

Based on long-term observations in Greece, Varotsos (2005) [23] has summarized his observation of SES (Seismic Electric Signal) activity; SES activity can frequently be detected in a short time interval of the order of one day. Large EQs then tend to take place about four weeks after such SES activity (*i.e.*, T = 4 weeks). With regard to the relationship of M *versus* D, Varotsos reported the following empirical relationship:

$$\text{Log (ED)} = a\,M + b \qquad (2)$$

Where E is the maximum amplitude of SES, and a and b are the constants determined empirically from the observational data. Equation (2) means that the SES intensity decreases with D as 1/D.

Compared to the lead time (T) of unusual animal behavior, the precursory time T of SES activity seems to be considerably larger as can be seen in Figure 3.

ULF/ELF Electromagnetic Emissions

This frequency range, especially ULF (f ≤ 1 Hz) has been extensively studied in different countries ever since three pioneering papers appeared for the Spitak [24], Loma Prieta [25] and Guam [26] EQs respectively. Hattori (2004) [27], Hayakawa and Hattori (2004) [28] and Molchanov and Hayakawa (2008) [14] summarized nearly all published data on seismogenic ULF emissions, plotted in Figure 4 in the form of D–M relations. The straight line, which marks the detection threshold of seismogenic ULF emissions (0.025 D = M – 4.5), is a linear regression line. According to these data, the maximum detection distance D is ~100 km for M = 7 and about 70–80 km for M = 6. The curve represents the empirical formula for abnormal animal behavior by Rikitake (1998) [4]. The straight line and the curve agree very closely for M values up to 7, but less so for the two events with M ≥ 7. This may simply be due to the fact that we used a linear regression line or to the fact that the number of events for M ≥ 7 is so small. When more events with M ≥ 7 become available, we will have to reexamine whether or not a present linear regression is acceptable.

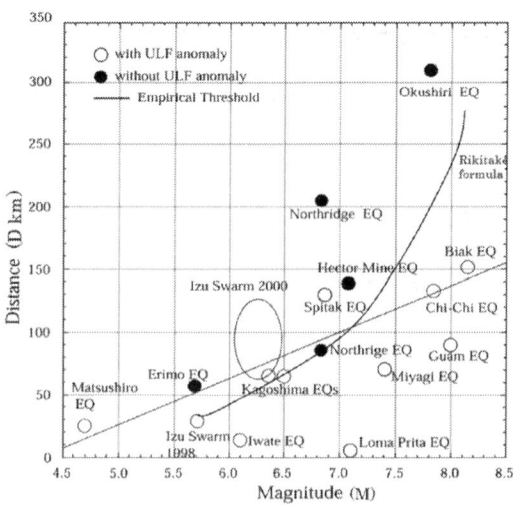

Figure 4: Summary of seismogenic ultra-low-frequency (ULF) radio emissions as M-D plot. Open circles indicate events with ULF anomaly, solid circles event without ULF anomaly. Straight line: empirical threshold (0.025 D =

M − 4.5) by linear regression. Curve: Rikitake's formula for unusual animal behavior.

Next we present the summary result of precursory time (T) of seismogenic ULF emissions based on the previous events as depicted in Figure 4 [29]. They show a typical temporal evolution.

- There seems to be no recognizable relationship between T and M.
- ULF emissions show first an intensity enhancement 1–2 weeks before an EQ, lasting for about one week (at least a few days). Then there is evidence for quiescence a few days before an EQ. A pronounced increase occurs a few days before the EQ, followed by an additional abrupt increase a few hours before the EQ.

A typical example (f = 0.01 Hz) of such a temporal evolution of seismogenic ULF emissions can be found for the Loma Prieta EQ [25]. The intensity of the first peak is 20 nT/sqrt(Hz), and the imminent peak amounted up to 60 nT/sqrt(Hz).

Let us compare these results with the corresponding animal behavior in the previous section. The precursory time, T does not seem to be dependent on the EQ M, and the above summary on the temporal evolution seems to be very consistent with that of unusual animal behavior with the first peak at about 7–10 days before an EQ and the imminent peak just before the EQ. There is a conspicuous quiescence between the two peaks in the case of seismogenic ULF emissions, which looks to be in agreement with Figure 3 for the animal behavior. Of course, we are not sure whether a minimum in the distribution of T in Figure 3 indicates a real quiescence in the temporal evolution of animal behavior for a particular event or if it is merely a statistical combination of two separate peaks. Figure 4 in the form of D *versus* M for seismogenic ULF emissions is found to be very similar to the empirical formula by Rikitake (1998) [4] for animal behavior alone, especially in the M range below M = 7. Unfortunately ULF emission reports for large to very large EQs (M = 7–8) are extremely rare, so it is not possible to make a relevant comparison of the two.

Next we discuss the summary of ELF seismogenic emissions in the frequency range of less than 10 Hz. There have been very few reports on the radio emissions in this frequency range [29]. According to [14] and [29], based on a few years observation in Kamchatka, it is found that the radio emission in the frequency range from a few to a few tens of Hz appears to occur a few days before an EQ. Though no M–D relationship data are available, the detection distance D was found to be less than 300 km.

ELF/VLF/LF Radio Noises

A significant number of papers have been published on radio emissions in these frequency ranges [30]. Once the radio emissions reach the atmosphere, they can propagate globally, with small attenuation, in the Earth-ionosphere waveguide (e.g., [31]). There have been very few reports on statistical studies on the characteristics of those seismogenic ELF/VLF radio emissions. Therefore, these frequency ranges are not quite suitable in the present paper with respect to a comparison with local unusual animal behavior.

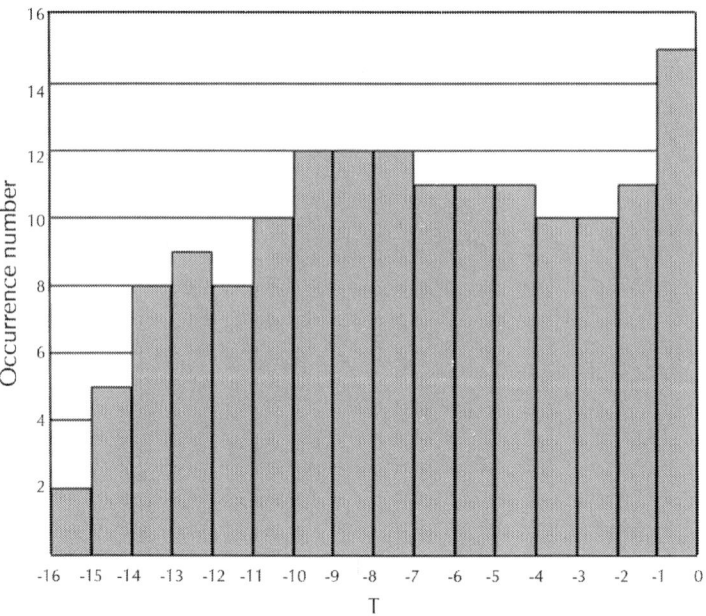

Figure 5: Occurrence histogram of pre-EQ extremely-low-frequency (ELF) radio emissions. Reproduced from a figure in Hata *et al.* (2006) [32] (with permission of the publisher) in which we define that strong emissions have a weight of unity and weak emissions have the weight of 0.5 (T in day).

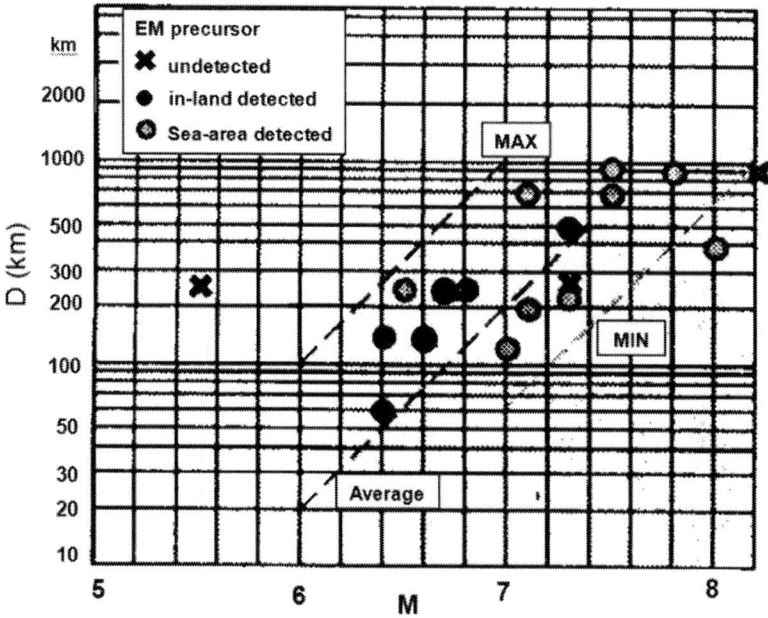

Figure 6: ELF radio noise plotted as M-D relation. Adapted from Hata *et al.* (2006) [32] with permission of the publisher.

However, there is one report on statistically relevant seismogenic ELF radio emissions (at 223 and 17 Hz) [32] drawing on observations over ten years. The precursory time (T) of these observations indicates that there is a peak in the occurrence 7–10 days before an EQ and an additional imminent peak just before the EQ (one day prior to the EQ), as shown in Figure 5. Figure 6 illustrates a statistical result on the M–D relationship, in which crosses indicate EQs with no ELF precursors, while the other symbols refer to EQs with ELF precursors. This figure shows that there is a general tendency that the detection distance (D) is larger for larger M values. The detection distance tends to be much larger for these frequencies than for the ULF emissions in Figure 4, due to the better propagation properties peculiar to this higher ELF range.

Additional studies on seismogenic VLF/LF radio emissions have been done [33,34]. However, the plots of M *versus* D and occurrence distribution of T are not available for comparison.

HF/VHF Radio Emissions

In this frequency range, we can expect the radio emissions to be spatially localized. Enomoto *et al.* (1999) [35] presented a statistical study on seismo-HF/VHF radio emissions on the basis of observations at Tsukuba, Japan over the course of a few years. The following characteristics were obtained: (1) HF/VHF radio emissions appeared within 3–4 days before an EQ and an abrupt increase occurred within one day before the EQ; (2) the detection distance was found to be several tens of km from the observatory, so that the HF/VHF radio emissions seem to be indeed highly localized; (3) no M-D information is available.

Seismo-Atmospheric and Seismo–Ionospheric Effects

Two reviews have been published, one on seismo-atmospheric effect [30] and the other on seismo-ionospheric perturbations [36]. The most efficient tool to study seismo-atmospheric perturbations is the use of the over-the-horizon VHF transmitter signals. It was found that abnormal reception of over-the-horizon VHF signals takes place about one week before an EQ. However, the mechanisms are uncertain as to how such atmospheric anomalies may be generated, even though there are a few possible mechanisms including ground surface temperature anomalies or positive hole effects [37]. In this sense, it is rather difficult to connect this phenomenon with unusual animal behavior.

Nearly the same situation holds for seismo-ionospheric perturbations. Hayakawa (2009) [36] has concluded that the lower ionospheric perturbation might be observed about one week before an EQ. While, Liu (2009) [38] indicated that the upper ionospheric (F region) perturbations take place a few days before the EQ. Again, the type of mechanisms involved in the generation of seismo-ionospheric perturbation is poorly understood at the moment, though a few hypotheses have been proposed [36] such as: (1) radon emanation and the associated electric field change [39]; (2) positive hole effects and the corresponding electric field generation [37]; and (3) atmospheric oscillation effect [36]. In either case of (1) or (2), the generation of electric field (DC) is essentially of importance, so that it may be related with the abnormal animal behavior. The third possibility relates to the precursory ground movement, which seems to be ruled out as in Section 2 in the sense of abnormal animal behavior.

DISCUSSIONS

Based on the suggestion of Rikitake (1998) [4], it appears that electromagnetic effects may be the most plausible candidate for causing abnormal animal behavior. In this paper, we first presented the statistical relations found by Rikitake between unusual behaviors of animals (dogs, cats, etc.) in terms of the three relationships of (i) M versus D, (ii) T versus M, and (iii) T occurrence histograms. Then we presented the statistical properties of seismogenic radio emissions in different frequency ranges while paying attention to the relationships (i)–(iii). Though the number of events for electromagnetic emissions is not large enough to have high statistical significance, it seems plausible as the result of comparisons, that the electromagnetic emissions in the ULF and lower ELF range are found to exhibit very similar characteristics in terms of the three relationships. Those emissions suggest a distinct temporal evolution: a first peak around one week before an EQ, followed by a second peak just before the EQ. This temporal change appears to be consistent with (or similar to) that of unusual animal behavior; the temporal evolution of unusual animal behavior also yields two peaks, a broad one about a week before an EQ and another just before the EQ.

In the field of experimental biology, a laboratory experiment has recently been attempted by Nishimura et al. (2010) [11], which appears to be worthwhile so as to understand our hypothesis presented in this paper. They have suggested that lizards are likely to perceive the low frequency electromagnetic signals. However, further expounded experiments are essential to have a better understanding of animal sensory perception because this kind of laboratory experiment is very time-consuming and obtaining any statistically reliable results is generally difficult.

It is further worthwhile to also mention studies of biological effects of radio signals by scientists in other disciplines. In the engineering

EMC (Electromagnetic Compatibility) area, great attention has been paid to the possible biological effects of electromagnetic radiation in different frequency ranges (from ELF (power line frequencies) to VHF or even higher (mobile phone frequencies)) [40]. Since Werthemier and Leeper (1979) [41] noted a higher incidence of cancer among children living in homes where ELF exposure was presumed to be higher than usual, there has been a very large number of studies in different countries on the biological effect of ELF power lines [42]. Even after such extensive investigations, a consensus seems not to have been reached yet regarding the effect of power lines on biology effects (human, animals, *etc*.). However, when looking at the complete list of papers in the summary report by [40] on: (1) the exposure of ELF magnetic field on animals; and (2) the relationship between ELF magnetic field exposure and cancers, it appears that a considerable number of papers suggest some influence on smaller animals (like mice or rats), while the relationship between ELF exposure and human cancer remains quite uncertain. The serious problem in the EMC area is the statistical reliability or significance of the data because the number of samples is generally not sufficient. This is nearly the same situation as in the study of macroscopic anomalies of EQs.

Next we discuss the biological effect of natural radio emissions. Cases of changes in the natural electromagnetic fields can be found in the scientific literature linking to observable effects on higher life forms which can also be found in the scientific literature. Such natural processes include solar, geomagnetic, cosmic ray, lightning activity, *etc*., and recently Cherry (2003) [43] suggested the importance of Schumann resonances in biology. Schumann resonances are a global resonance phenomenon excited primarily by background lightning discharges in the Earth-ionosphere waveguide [31]. The Schumann resonances are very weak, but very stationary with distinct frequencies at 8, 14, 20 Hz, *etc*. The high stationarity of Schumann resonances stands in sharp contrast to previously mentioned natural phenomena which are very transient. We have studied the biological effect of this Schumann resonance on the basis of our own ELF observation in Moshiri, Hokkaido and the simultaneous observation of human blood pressure, heart rate and depression, *etc*. [44]. These data were obtained between April and July 2001. It was found that the blood pressure in humans shows statistically significant mean differences between normal and enhanced Schumann resonance days. That is, the mean blood pressure rate is significantly lower for enhanced Schumann resonance days than for normal Schumann resonance days. The ELF magnetic field of Schumann resonances is extremely weak (<1 pT/sqrt(Hz)), so that its stationarity (or persistence) appears to be of primary importance.

Finally, we suggest that any ULF/ELF seismogenic radio emissions may be a dominant source of unusual animal behavior before an EQ. The intensity of seismogenic ULF/ELF emissions is on the order of 1–50 nT (1 nT corresponds to 0.3 V/m in the atmosphere). These values have been used in the EMC area, to derive, for the purpose of risk management, a maximum permissible exposure, i.e. the field intensity (either electric or magnetic) multiplied by the exposure time is thought to be the primary factor in studies of abnormal animal behavior. In other words, even though the field intensity may not be very large (such as seismogenic noises (Schumann resonance as well)), the persistence or prolongation of the radiation may play an essential role in animal behavior. For example, seismogenic precursory ULF emissions are known to persist, at least, for a few days or even up to about one week. Of course, there remain so many questions; e.g., is the electric or magnetic field influential on the animals?

CONCLUSIONS

Based on the intense comparison of characteristics of abnormal animal behavior and seismogenic electromagnetic radiation in a wide frequency range, we come to the conclusion that lower frequency (such as ULF and ELF) electromagnetic emissions are a plausible candidate to explain abnormal animal behavior before an EQ. In order to verify this hypothesis, the following steps are essential: (i) further anecdotal and retrospective studies of abnormal animal behavior; (ii) a coordinated measurement of animal behavior with seismic and chemical sensors in combination with electromagnetic sensors in a seismically active region (as suggested in [5]).

REFERENCES AND NOTES

1.   Evernden, J. U.S. Geological Survey Office of Earthquake Studies. In *Abnormal Animal Behavior Prior to Earthquakes*;

U.S. Department of Commerce, National Technical Information Service: Alexandria, VA, USA, 1976.

2.   Buskirk, R.E.; Frohlich, C.; Lantham, G.V. Unusual animal behavior before earthquakes: A review of possible sensory mechanisms. *Rev. Geophys. Space Phys.* 1981, *19*, 247–270.

3.   Tributsch, H. *When the Snakes Awake—Animals and Earthquake Prediction*; MIT Press: Cambridge, MA, USA, 1982.

4.   Rikitake, T. *The Science of Macro-Anomaly Precursory to an Earthquake*; Kinmiraisha: Nagoya, Japan, 1998.

5.   Kirschvink, J.L. Earthquake prediction by animals: Evolution and sensory perception. *Bull. Seismol. Soc. Am.* 2000, *90*, 312–323.

6.   Ikeya, M. *Earthquakes and Animals: From Folk Legends to Science*; World Scientific: Singapore, 2004.

7.   Yokoi, S.; Ikeya, M.; Yagi, T.; Nagai, K. Mouse circadian rhythm before the Kobe earthquake in 1995. *Bioelectromagnetics* 2003, *24*, 289–291.

8.   Li, Y.; Liu, Y.; Jiang, Z.; Guan, J.; Yi, G.; Cheng, S.; Yang, B.; Fu, T.; Wang, Z. Behavioral change related to Wenchuan devastating earthquake in mice. *Bioelectromagnetics* 2009, *30*, 613–620.

9.   Bhargava, N.; Katiyar, V.K.; Sharma, M.L.; Pradhan, P. Earthquake prediction through animal behavior: A review. *Indian J. Biomech.* 2009, *7–8*, 159–165.

10.   Grant, R.A.; Halliday, T. Predicting the unpredictable; evidence of pre-seismic anticipatory behavior in the common toad. *J. Zool.* 2010, *281*, 1–9.

11.   Nishimura, T.; Okano, H.; Tada, H.; Nishimura, E.; Sugimoto, K.; Mohri, K.; Fukushima, M. Lizards respond to an extremely low-frequency electromagnetic field. *J. Exp. Biol.* 2010, *213*, 1985–1990.

12.   Grant, R.A.; Halliday, T.; Balderer, W.P.; Leuenberger, F.; Newcomer, M.; CYR, G.; Freund, F.T. Ground water chemistry changes before major earthquakes and possible effects on animals. *Int. J. Environ. Res. Public Health* 2011, *8*, 1936–1959.

13.   *Seismo Electromagnetics: Lithosphere-Atmosphere-Ionosphere Coupling*; Hayakawa, M., Molchanov, O.A., Eds.; TERRAPUB: Tokyo, Japan, 2002.

14. Molchanov, O.A.; Hayakawa, M. *Seismo Electromagnetics and Related Phenomena: History and Latest Results*; TERRAPUB: Tokyo, Japan, 2008.

15. Kayano, I. Damages and seismicity distribution of an earthquake (M6.0) occurred in the Ibaraki-ken Nambu region on February 27, 1983 (in Japanese). *Mem. Earthquake Res. Inst.* 1983, *58*, 831–878.

16. Kayano, I. *Study of Nagano-ken Seibu Earthquake (in Japanese); Technical Report*; Natural Disaster Science Study Group: Tokyo Japan, 1984; pp. 135–143.

17. *Atmospheric and Ionospheric Electromagnetic Phenomena Associated with Earthquakes*; Hayakawa, M., Ed.; TERRAPUB: Tokyo, Japan, 1999.

18. Pulinets, S.A.; Boyarchuk, K. *Ionospheric Precursors of Earthquakes*; Springer: Berlin, Germany, 2004.

19. *Electromagnetic Phenomena Associated with Earthquakes*; Hayakawa, M., Ed.; Transworld Research Network: Trivandrum, India, 2009.

20. *The Frontier of Earthquake Prediction Studies*; Hayakawa, M., Ed.; Nihon-senmontosho-Shuppan: Tokyo, Japan, 2012.

21. Uyeda, S.; Nagao, T.; Kamogawa, M. Short-term earthquake prediction: Current state of seismo-electromagnetics. *Tectonophysics* 2009, *470*, 205–213.

22. Hayakawa, M.; Hobara, Y. Current status of seismo-electromagnetics for short-term earthquake prediction. *Geomat. Nat. Hazardz Risk* 2010, *1*, 115–155.

23. Varotsos, P. *The Physics of Seismic Electric Signals*; TERRAPUB: Tokyo, Japan, 2005.

24. Kopytenko, Y.A.; Matiashvili, T.G.; Voronov, P.M.; Kopytenko, E.A.; Molchanov, O.A. Detection of ultra-low frequency emissions connected with the Spitak earthquake and its aftershock activity based on geomagnetic pulsations data at Dusheti and Vardzia observations. *Phys. Earth Planet. Inter.* 1993, *77*, 85–95.

25. Fraser-Smith, A.C.; Bernardi, A.; McGill, P.R.; Ladd, M.E.; Helliwell, R.A.; Villard, O.G., Jr. Low-frequency magnetic field measurements near the epicenter of the Ms7.1 Loma Prieta earthquake. *Geophys. Res. Lett.* 1990, *17*, 1465–1468.

26. Hayakawa, M.; Kawate, R.; Molchanov, O.A.; Yumoto, K. Results of ultra-low-frequency magnetic field measurements during the Guam earthquake of 8 August 1993. *Geophys. Res. Lett.* 1996, *23*, 241–244.

27. Hattori, K. ULF geomagnetic changes associated with large earthquakes. *Terr. Atmos. Ocean. Sci.* 2004, *15*, 329–360.

28. Hayakawa, M.; Hattori, K. Ultra-low-frequency electromagnetic emissions associated with earthquakes. *IEEJ Trans. Fundament. Mater.* 2004, *124*, 1101–1108.

29. Schekotov, A.Y.; Molchanov, O.A.; Hayakawa, M.; Fedorov, E.N.; Chebrov, V.N.; Sinitsin, V.I.; Gordeev, E.E.; Belyaev, G.G.; Yagova, N.V. ULF/ELF magnetic field variations from atmosphere induced by seismicity. *Radio Sci.* 2007, *42*.

30. Hayakawa, M. Seismogenic perturbation in the atmosphere. In *Electromagnetic Phenomena Associated with Earthquakes*; Hayakawa, M., Ed.; Transworld Research Network: Trivandrum, India, 2009; pp. 119–136.

31. Nickolaenko, A.P.; Hayakawa, M. *Resonances in the Earth-Ionosphere Cavity*; Kluwer Academic Publishers: Dordrecht, The Netherlands, 2002.

32. Hata, M.; Takumi, I.; Yasukawa, H.; Fujii, T. ELF band EM precursor and signal processing to predict earthquakes. In *Natural Electromagnetic Phenomena and Electromagnetic Theory*; The Institute of Electrical Engineers of Japan: Tokyo, Japan, 2006; pp. 46–52.

33. Fujinawa, Y.; Takahashi, K.; Matsumoto, T.; Kawakami, N. Sources of earthquake-related VLF electromagnetic signals. In *Atmospheric and Ionospheric Electromagnetic Phenomena Associated with Earthquakes*; Hayakawa, M., Ed.; TERRAPUB: Tokyo, Japan, 1999; pp. 405–415.

34. Oike, K.; Yamada, T. Relationship between shallow earthquakes and electromagnetic noises in the LF and VLF range. In *Electromagnetic Phenomena Related to Earthquake Prediction*; Hayakawa, M., Fujinawa, Y., Eds.; TERRAPUB: Tokyo, Japan, 1994; pp. 115–130.

35. Enomoto, Y.; Hashimoto, H. Anomalous electric signals detected before recent earthquakes in Japan near Tsukuba. In

Electromagnetic Phenomena Related to Earthquake Prediction; Hayakawa, M., Fujinawa, Y., Eds.; TERRAPUB: Tokyo, Japan, 1994; pp. 261–269.

36. Hayakawa, M. Lower ionospheric perturbation associated with earthquakes, as detected by subionospheric VLF/LF radio waves. In *Electromagnetic Phenomena Associated with Earthquakes*; Hayakawa, M., Ed.; Transworld Research Network: Trivandrum, India, 2009; pp. 137–185.

37. Freund, F. Stress-activated positive hole change carriers in rocks and the generator of pre-earthquake signals. In *Electromagnetic Phenomena Associated with Earthquakes*; Hayakawa, M., Ed.; Transworld Research Network: Trivandrum, India, 2009; pp. 41–96.

38. Liu, J.Y. Earthquake precursors observed in the ionospheric F-region. In *Electromagnetic Phenomena Associated with Earthquakes*; Hayakawa, M., Ed.; Transworld Research Network: Trivandrum, India, 2009; pp. 187–204.

39. Pulinets, S.A. Lithosphere-atmosphere-ionosphere coupling (LAIC) model. In *Electromagnetic Phenomena Associated with Earthquakes*; Hayakawa, M., Ed.; Transworld Research Network: Trivandrum, India, 2009; pp. 235–253.

40. *Standard for Safety Levels with Respect to Human Exposure to Radio Frequency Electromagnetic Fields, 3 kHz to 300 GHz; IEEE Std C95.1, 1999*; IEEE: New York, NY, USA, 1999.

41. Werthemier, N.; Leeper, E. Electrical wiring configuration and childhood cancer. *Am. J. Epidemiol.* 1979, *109*, 273–248.

42. Reiter, R.J. An assessment of the bioeffects induced by power-line frequency electromagnetic fields. In *Modern Radio Science 1999*; Stuchly, M., Ed.; Oxford University Press: Oxford, UK, 1999; pp. 287–307.

43. Cherry, N.J. Human intelligence: The brain, an electromagnetic system synchronized by the Schumann resonance signal. *Med. Hypoth.* 2003, *60*, 843–844.

44. Mitsutake, G.; Otsuka, K.; Hayakawa, M.; Sekiguchi, M.; Cornélissen, G.; Halberg, F. Does Schumann resonance affect our blood pressure? *Biomed. Pharmacother.* 2005, *59*, S10–S14.

Chapter 4

A Deterministic Approach to Earthquake Prediction

Vittorio Sgrigna[1] and Livio Conti[2]

[1]Dipartimento di Fisica and Sezione INFN, Università Roma Tre, 84 Via della Vasca Navale, 00146 Roma, Italy
[2]Facoltà di Ingegneria, Università Telematica Internazionale UNINETTUNO, Corso Vittorio Emanuele II 39, 00186 Roma, Italy

ABSTRACT

The paper aims at giving suggestions for a deterministic approach to investigate possible earthquake prediction and warning. A fundamental contribution can come by observations and physical modeling of earthquake precursors aiming at seeing in perspective the phenomenon earthquake within the framework of a unified theory able to explain the causes of its genesis, and the dynamics, rheology, and microphysics of its preparation, occurrence, postseismic relaxation, and interseismic phases. Studies based on combined ground and space observations of earthquake precursors are essential to address the issue. Unfortunately, up to now, what is lacking is the demonstration of a causal relationship (with explained physical processes and looking for a correlation) between data gathered simultaneously and continuously by space observations and ground-based measurements. In doing this, modern and/or new methods and technologies have to be adopted to try to solve the problem. Coordinated space- and ground-based observations imply available test sites on the Earth surface to correlate ground data, collected by appropriate networks of instruments, with space ones detected on board of Low-Earth-Orbit (LEO) satellites. Moreover, a new strong theoretical scientific effort is necessary to try to understand the physics of the earthquake.

INTRODUCTION

In our opinion, the investigation of possible earthquake prediction must be carried out on a deterministic basis. Unfortunately, at the moment, the study of the physical conditions that give rise to an earthquake and the processes that precede a seismic rupture of an ordinary event are at a very preliminary stage and, consequently, the techniques of prediction of time of origin, epicentre, and magnitude of an impending earthquake now available are below standard.

Therefore, the present level of knowledge is unable to achieve the objective of a deterministic prediction of an ordinary seismic event, but it certainly will in a more or less distant future tackle the problem with seriousness and avoiding scientifically incorrect, wasteful, and inconclusive shortcuts, as sometimes has been done. It will take long time (may be years, tens of years, or centuries) because this approach requires a great cultural, financial, and organizational effort on an international basis. It implies the need for carrying out combined ground and near-Earth space continuous observations of the so-called earthquake precursors, coseismic and postseismic phenomena, as well as the development of appropriate theoretical models able to justify the observations in order to understand the physical mechanisms underlying the earthquake preparation and occurrence. So, ground networks of instruments in the major seismic areas of the Earth and Low-Earth-orbit (LEO) multi-instrument satellites, as well as laboratory and theoretical investigations, will be necessary to address the study carried out by coordinate teams of researchers and specialists in the different scientific and technical fields of the physics of the Earth system. Probably, the pressure of act more quickly sometimes gives bad advise. An example of such behaviour has been given even on the occasion of the recent destructive seismic event occurred in Japan last March 2011 when, also inside groups of the scientific community, reckless statements were raised hinting the hypothesis (and someone has actually said) that earthquake prediction is possible, especially if it is possible there will be financial support and some kind of scientific coordination.

Remember that the 2011 March 11 05:46:23 UTC Tohoku earthquake (near the East coast of Honshu, Japan), also knows as the Great East Japan Earthquake, had magnitude 9.0; location 38.322°N, 142.369°E; depth 32 km. It was the Japan's most powerful earthquake since records began and has struck the north-east coast, triggering a massive tsunami. The Japanese National Police Agency has confirmed 15.550 deaths, 5.688 injured, and 5.344 people missing across twenty-tow prefectures, as well as about 225.000 buildings damaged or destroyed [1].

In the face of such huge disaster, the above-mentioned claims on the earthquake prediction must be considered as regrettable. They were issued through mass media and even within a pseudoscientific context, and appearing as a kind of "scientific looting." Such false statements can only be used to take advantage of the disaster, maybe to obtain more easily research funds or for a greater visibility within the scientific community, civil services, and authorities that need to take adequate measures for assistance and protection of the population and reconstruction of houses and infrastructures. To justify the concept of earthquake prediction, "noises" are often introduced thus confusing different concepts such as earthquake precursor, seismic hazard, earthquake warning, and earthquake forecasting. A similar disgraceful behaviour does not produce any result useful to science or to society.

This "vulnus" inside the scientific community cannot easily be healed and overcame, since mediocre minds are as able to organize themselves as brilliant ones. So, self-referential poor groups of researchers are easily formed and can also permeate international peer-review systems.

But any honest scientist knows that the way to go is almost always one more long and tiring. It requires intelligence, time, perseverance, and scientific humility and honesty.

As mentioned above, a possible contribution to a deterministic earthquake prediction approach may be given by observations and physical modelling of earthquake precursors aimed at seeing, in perspective, the earthquake phenomenon within the framework of a unified theory able to explain the causes of its genesis, and the dynamics, rheology, and microphysics of its preparation, occurrence, postseismic relaxation, and interseismic phases. Unfortunately, up to now what is lacking is the demonstration of a causal relationship (with explained physical processes and looking for a correlation) between data gathered simultaneously and continuously by space observations and ground-based measurements. In doing this, modern and/or new methods and technologies have to be adopted to try to solve the problem.

Within this framework, a few projects and experiments have been carried out on the subject by our team and accompanied by specific theoretical interpretations. They are reported in the paper. As an introduction and justification to these studies and also to avoid confusion, we try to clarify some basic concepts on the matter, critical and methodological aspects concerning deterministic and statistic approaches, and their use in earthquake prediction and warning.

The earthquake prediction and damage prevention methods, as well as the analysis of lithosphere-atmosphere couplings associated with the preparation of seismic events, are the introductory and basic elements of the paper. They will be discussed in this section.

Earthquake Damage Prevention and Deterministic Prediction Concepts

It is well known that earthquakes are a manifestation of significant ground rock deformation events, that is, episodic deformations of the upper and, more or less, brittle layers of the Earth's lithosphere.

These can be classified as fast seismic ruptures, slow earthquakes, and subseismic events. Since the energy released during large earthquakes affects human life, the development and application of appropriate and efficient techniques to defend society from these destructive effects are necessary. At the present time, only two suitable approaches are available: damage prevention and prediction methods.

Earthquake damage prevention implies the development of methods for evaluating seismic risks in order to enable disaster assessment and techniques for use in estimating seismic risk, with the ultimate aim of reducing damage produced by earthquakes through reliable means. The prevention of damage is achievable with existing state of knowledge. In this approach, a great importance lies in the optimization of methods necessary to determine the three main factors—vulnerability, value, and hazard—which define seismic risk.

In contrast, the deterministic prediction of the time of origin, hypocentral (or epicentral) location, and magnitude of an impending earthquake is an open scientific problem. The reason for this is that such predictions are based on the detection of the so-called earthquake precursors or pre-earthquake phenomena, and the physical interpretation of these is a very complicated matter.

At this point, a few main concepts on precursor detectability must be considered. First, it must be clear that reducing "physics of the earthquake" only to the creation of fault rupture and consequent seismic wave propagation is to oversimplify the problem. In fact, it has been repeatedly observed that part of the accumulated (preseismic) elastic energy is also converted to other kind of energies (electromagnetic, acoustic, heat, etc.) and that these conversion mechanisms are probably similar to that of seismic energy. Moreover, observations during interseismic and preseismic periods indicate that large earthquakes are often preceded by signals of different natures (the so-called earthquake precursors), of which the mechanical (tilt and strain), gaseous (helium and radon), and electromagnetic ones have been demonstrated to be the most significant manifestations (see this paper and also [2]). However, the study of the physical conditions that give rise to an earthquake and of the processes that precede a seismic rupture is at a very preliminary stage and, consequently, the techniques of prediction available at the moment are below standard.

In trying to by-pass these difficulties, many investigators have likely been attracted by a statistical prediction approach based on the so-called earthquake forecasting method, that is, the probability of occurrence of an event in a given geographical location, within assigned values of magnitude and time ranges. However, even though the forecasting methods, such as those of the M8 and CN algorithms (e.g., [3–5]) or of the acceleration deformation approach (e.g., [6]) have reached a very good level of maturity and can display a good level of importance and practical use, they overlap with the seismic hazard concept, one of the three factors used to estimate seismic risk. This could result in a possible ambiguity in the application of earthquake prediction and earthquake damage prevention approaches, which could give rise to a kind of "methodological noise" that would be capable of introducing systematic errors in the use of the two methods. We, therefore, believe that it should be better to pursue the deterministic prediction approach even if a reliable deterministic method of earthquake prediction will presumably be available only in the more distant future.

This conclusion is also confirmed by the underestimated expectation of earthquake prediction in a relatively short period of time based on the basis of seismic precursor studies carried out in the last decades. As mentioned above, the physics of earthquakes has been demonstrated to be a very complicated matter. Nevertheless, research with this aim continues with a critical view, new ideas, and thorough investigations, and the results seem to be promising. Therefore, we propose to carry out studies based on the physics of earthquake precursors, including the necessary field measurements in seismic areas and appropriate laboratory and theoretical investigations to corroborate the observations.

Progress in this area could be due not only to increased amounts and accuracy of ground field measurements, careful attention to errors in data, and improved understanding of earthquake source mechanics, but also—and possibly most importantly—to a new approach based on observations from space.

But how to reach a deterministic seismic prediction by earthquake precursors needs to be better clarified since it is considered by several authors that such an approach seems to be unadvisable because for a deterministic prediction the space localization (epicentre or hypocenter), the time of origin, and the energy or magnitude of an impending earthquake are required at the same time. A possible method on how in principle to practically predict earthquakes with precursory phenomena is proposed at the beginning of Section 3.

Seismoelectromagnetic Emissions and Couplings between Solid Earth and Near-earth Space

A great contribution for constructing a deterministic prediction model arises by pre-earthquake (or precursory) phenomena, since they may help in understanding the physical mechanisms underlying the preparation phase of a seismic event. It has been shown that in the Earth's crust, rock microfracturing preceding a seismic rupture may cause local surface deformation fields, rock dislocations, charged particle generation and motion, electrical conductivity changes, gas emission, fluid diffusion, electrokinetic (EKE), piezomagnetic, and piezoelectric effects. It has also been proposed that charge carriers could be activated in dry rocks mainly by the increasing external stress. These mechanisms have been considered as the main sources of the so-called seismo electro -magnetic emissions (SEME) consisting of broad-band (from DC to a few tens of MHz) electromagnetic (EM) fields observed at the Earth's surface and in the near-Earth space (neutral and ionised atmosphere and magnetosphere). Electromagnetic emissions (EMEs) radiated from the Earth's surface and produced as a consequence of earthquake preparation and occurrence, or by human activities, demonstrated to propagate through the near-Earth space and to cause perturbations of electric and magnetic fields and Van Allen radiation belt particle precipitations, ionospheric variations of temperature and density of the ionic, and electronic plasma components in the topside ionosphere. These perturbations are detectable by Low-Earth-Orbit (LEO) satellites [2, 7–9].

Within this framework, natural disasters, such as earthquakes, and the impact of anthropogenic EME waves (power line harmonic radiation, VLF transmitters, HF broadcasting stations) in the near-Earth space can also be considered as coupling elements of the lithosphere-atmosphere-ionosphere-magnetosphere interactions. All above mentioned suggests that to better investigate the phenomenon earthquake, simultaneous and coordinated space and ground-based observations in seismic areas have to be carried out. The main problem in this studies is to reconcile near-Earth space perturbations only with the propagation of SEME-waves through the atmosphere and magnetosphere, filtering from the data the impact of atmospheric EME waves during thunderstorm activity, and effects of sun and cosmic rays in the geomagnetic cavity.

Space observations are being performed or are going to be carried out, in the ionosphere-magnetosphere transition region, and a few satellite missions (Demeter, QuakeSat, Sich-1 M, Compass-1/2, Esperia, Egle, Arina, Ausonia, etc.) have already been carried out and/or are proposed from 2001 until the present [2, 10–15].

The basic premise is that observations of different ground and space seismic precursors as well as laboratory experiments on rocks and the development of theoretical models, all of which aimed at placing the phenomenon "earthquake" within the framework of a unified theory, would be able to explain the causes of its genesis, and the dynamics, rheology, and microphysics of its preparation, occurrence, postseismic relaxation and interseismic phases. The physical system to be considered includes solid Earth and nearEarth space with related couplings and perturbations. Also, it is hoped that a better scientific coordination on an international basis between diverse teams of researchers would smooth out and integrate different methodological approaches relatively to each other for a better use of the different competences, instruments, and databases. Up to now what is lacking is the demonstration of a causal relationship with explained physical processes and looking for a correlation between data gathered simultaneously and continuously by space observations and ground-based measurements. That is why we believe that the best approach is to plan and design coordinated and simultaneous ground-based measurements (carried out by appropriate networks of instruments in available test sites on the Earth surface) to be correlated with multiparametric space observations onboard satellites, together with the development of appropriate methods of data analysis and theoretical modeling. To this end, we have installed the TELLUS tilt network in the seismic area of the Central Apennines of Italy. This network will, in the near future, also include magnetometers and specific devices to detect electric field. Results obtained by the TELLUS network have been reported [16]. Within the framework of a guest investigation programme we have studied data collected in the topside ionosphere by the DEMETER microsatellite, proposed a specific LEO satellite project (ESPERIA) and built and tested in space two ESPERIA instruments (the EGLE magnetometer and ARINA particle detector). At the same time, we also have made first attempts to develop a theoretical model of the genesis and propagation of preearthquake electromagnetic emissions in the lithosphere and near-Earth space [7, 15, 17–19].

In 2007, after an IUGG resolution in support of ESPERIA (2007 IUGG resolution number 5) for an ionospheric space mission, we submitted to the Italian Space Agency (ASI) a new space project (AUSONIA), with more large scientific objectives than those of ESPERIA. AUSONIA includes the monitoring and mapping of the ionosphere and of the Earth magnetic field and also the study of tropospheric transient emissions [14]. Then, AUSONIA can investigate both perturbative and steady-state phenomena.

Next two sections will clarify basic concepts concerning hypocentral focal zone and epicentral precursory area (Section 2) and refer to reliable results reported in literature about earthquake precursors (Section 3) and their possible use as seismic predictors. The following Sections 4-5 report the ESPERIA and AUSONIA space mission projects and the description and testing of the first ESPERIA and AUSONIA instruments: the EGLE magnetometer and ARINA particle detector.

HYPOCENTRAL PREPARATION FOCAL ZONE AND EPICENTRAL PRECURSORY AREA

The most familiar brittle lithospheric deformation event is defined as ordinary earthquake, that is, a deformation, fracture, structure, and phase transformation phenomenon, which releases suddenly a large amount of the elastic energy stored in the medium and is accompanied by a substantial fraction of energy radiated as elastic (seismic) waves. Seismic wave energy is a certain part (from about 1 to 10%) of total (radiated and not radiated) energy, and it is usually assumed as an estimate of the total energy of the earthquake. Moderate and strong earthquakes, with magnitude from 5.0 to 9.0, have energy and seismic moment [20] approximately in the range $10^{12}-10^{18}$ J and $10^{17}-10^{22}$ Nm, respectively, as given by the following well-known relationships (in cgs units) between energy (E), scalar seismic moment (M_0), and surface earthquake magnitude (M_s):

$$\log E = 11.8 + 1.5M_S,$$
$$\log M_0 = 1.5M_S + 16.1. \tag{1}$$

But reducing "physics of the earthquake" only to the creation of fault rupture and consequent seismic wave radiation is to oversimplify the problem. It has been repeatedly observed that part of the accumulatedpreseismic elastic energy is also converted to other kind of energies (electromagnetic and acoustic ones, heat, etc.) and these (yet unknown) conversion mechanisms are probably similar as that to seismic energy. The understanding of such preseismic processes is fundamental to plan and design earthquake prediction techniques on a deterministic basis, that is, based on the so-called seismo-associated phenomena or earthquake precursors. The latter are phenomena of different types (seismic and nonseismic ones) accompanying the characteristic deformation of rocks during earthquake preparation time or preseismic period, and associated with changes in physical conditions in the so-called preparation focal zone (volume) as defined by standarddilatancy-diffusion and crack-avalanche "dilatancy" models [21–23].

Until now, no exhaustive physical models have been proposed and accepted by the scientific community to be used for a deterministic earthquake prediction approach. What is known on the topic is that in the time interval preceding a seismic fracture, stress and strain energy are accumulated in a fault asperity. Most of investigators consider reasonable to assume this increasing and concentrating stress at depth as a cause of the anelastic volumetric increase (dilatancy) of a relatively small portion of rock, and consequent rock dislocation and microfracturing. This volume of cracked rock at depth (preparation focal zone) is considered as a primary local source of precursor signals. These signals propagating in the surrounding medium allow the earthquake precursors to be observed in a finite region of the Earth's surface (precursor area). Then, in principle earthquake precursors can be used to indicate the impending occurrence of a seismic event. Characteristic sizes of the preparation focal zone and of the precursor area have been estimated by Dobrovolsky et al. [24, 25]. They found the volume (V) of soft inclusion (cracked rock) at depth in the lithosphere versus magnitude (M), is described as follows:

$$V_{max} = 10^{(1.24M-4.47)} \text{ km}^3,$$

(2)

which for a spherical volume of radius (r) gives:

$$r = 10^{0.414M-1.696} \text{ km.}$$

(3)

The dimension of the precursor region at the earth surface is defined [24] by the radius (R) of the Earth's surface area where preseismic strain changes exceed tidal strains ($\approx 10^{-8}$), as follows

$$R = 10^{0.43M} \text{ km.}$$

(4)

Relationships between preseismic strain ε, magnitude M, and distance R are

$$\varepsilon = \frac{10^{1.5M-9.18}}{R^3} \quad \text{for } M < 5.0,$$

$$\varepsilon = \frac{10^{1.3M-8.19}}{R^3} \quad \text{for } M \geq 5.0.$$

(5)

For comparison, we report in Table 1 the characteristic dimensions of the preparation focal zone at depth (i.e., the source of earthquake precursors) with those of the precursor region at the Earth's surface. Data are obtained for $4.0 \leq M \leq 7.0$ events, in the simple case of a preparation focal area modelled by a spherical volume (v) and in presence of a homogeneous medium.

Table 1: Sizes of earthquake preparation zone (r) and precursor region (R) for $4.0 \leq M \leq 7.0$

M	r (km)	R (km)
4.0	0.1	52
5.0	2.5	141
6.0	6.0	380
7.0	41.3	1023

It can be seen that by basing on the model by Dobrovolsky et al. [24] characteristic sizes of preparation focal area at depth are relatively small (from a few hundred meters to a few ten of kilometres) when compared with those of the precursor region at the Earth's surface (from a few tens of kilometres to about one thousand of kilometres).

We stress that this result is only valid for local deformation (tilt and strain) and for a homogeneous Earth's crust containing a soft inclusion simulating the rheological behaviour of a preseismic dilatants volume of cracked rock. When a more realistic and complicated geometry and structure is assumed for the Earth's crust in a seismic region and/or when other kinds of earthquake precursors than mechanical ones are considered (for instance electric and magnetic fields), a new general and more specific physical model must be proposed to determine the above-mentioned r and R sizes of the preparation focal zone and precursor area. In particular, the presence of discrete geodynamic structures (crustal blocks) in seismic regions (see, [16, 26]) implies that a preseismic deformation (tilt and strain) field may propagate at different distances and velocities in the different directions from the preparation focal area. This anisotropic space and time distribution of the preseismic deformation field mainly depends on dimensions, geometry, structure, and rheology of crustal blocks and their transition zones [7, 17].

Finally, empirical semilogarithmic relationships have also been proposed by several authors between magnitude M of an impending earthquake and precursory time ΔT (interval between the onset time of a precursor signal and the time origin of the earthquake). One of such relationships proposed by Rikitake [27] is

$$\log \Delta T = 0.76 M - 1.83.$$

(6)

Concerning nonmechanical earthquake precursors, a model of preseismic electromagnetic emissions is in preparation, which first results have been reported in international meetings [7, 12, 28].

MORE RELIABLE GROUND AND SPACE EARTHQUAKE PRECURSORS

In general, earthquake precursors can be divided in the two classes of so-called seismic and nonseismic phenomena. In the class of seismic phenomena are included seismic gap, decreasing (seismic quiescence) and increasing background seismicity, and change in the seismic wave velocity. The list of nonseismic phenomena includes numerous earthquake precursors of very different types as phenomena directly reconciled with local deformations (ground elevations and tilts, strains in rock, water levels in wells, etc.) or of other kind as electric and magnetic fields, EM emissions, electric resistivity in rock, acoustic emissions, gas exhalations (mainly radon and helium), and so forth. The time scale of an earthquake prediction attempt is by convention generally classified as short term (≈hour–days), long-term (≈years–decades), intermediate-term (≈weeks–years), according to the expected time interval to the earthquake (precursor time). Really, only short-term and intermediate term time scales can be considered for a true deterministic earthquake prediction methods, since long-term one, in practice, can be identified with the seismic assessment of the seismic hazard of a given zone and, then, associated with the statistical probability for the occurrence of large earthquakes.

A Possible First Empirical Approach to Deterministic Earthquake Prediction Based on Precursory Phenomena

A deterministic earthquake prediction method based on precursory phenomena has not yet been proposed. At this purpose, the combination of simultaneous and continuous observations of mechanical medium-term precursors and electromagnetic short-term ones in selected seismic test areas could be of particular importance in determining, within the time interval of the short-term precursory time (hour-days), the epicentre, the magnitude and time of occurrence of an impending earthquake.

In principle, as a first empirical approach, a possible method could be to combine the most reliable medium-term and short-term earthquake precursors, as follows.

- *First Warning/Alert*

One could imagine using the onset times of the anomalous medium-term (weeks-months) tilt and strain signals recorded by the multi-instrument network working in the seismic test area as a first-time warning.

- Second (Final) Warning/Alert

A second (final) time warning could be associated with the onset times of the first anomalous short-term (hour-days) electromagnetic signals pointed out by the same instrumental network. Then, the uncertainty in the estimate of time origin of the event will be ranging from ~1 hour to days.

- Epicenter/Hypocenter

An estimate of the future epicenter could be attempted by the time shifts between the onset times of the different medium-term anomalous mechanical signals observed by the instrumental network and on the basis of the velocity of propagation of the preseismic deformation front through the crust block structures of the observed test area. This velocity is calculated to be of the order of 1 cm/s [16, 29–31]. But this value must be determined for each test area.

- Magnitude

Finally, the magnitude could be roughly estimated on the basis of the empirical relationships between magnitude and precursory time (e.g., (6) in Section 2) by the time shifts between the onset times of all the couple of medium-term and short-term signals observed at each site of the instrumental network. The use of amplitudes of such signals to calculate the magnitude appear to be more questionable since some spatial differential amplification effects (site effects) are observed in the different sites (then, in the different blocks) where instruments are located. An example of such site effect can be observed in Figure 1.

But before applying, any deterministic method of prediction quantitative specific physical models, unavailable at moment, must be proposed for each test area in order to describe the geodynamics and rheology of crustal blocks and relative transition zones, as well as the physical mechanisms underlying the mechanical and electromagnetic preseismic sources. In particular, to justify the observations is necessary to model the shape, onset times, and durations of precursory signals, thus, reconciling them with the preseismic source behaviour and characteristics (space localization, dimensions, geometry, space orientation, rock yielding conditions, and catastrophic rupture mechanisms).

Only at this stage, an exhaustive and general physical interpretation of such precursors could be of help in reducing the uncertainty (physical error) in the estimation of the epicentral position, magnitude, and time of origin of an impending earthquake, then in contributing to define an acceptable deterministic earthquake prediction. Up to now, there have been systematic observations of mechanical intermediate-term and electromagnetic short-term precursors, which have been shown to be more suitable for the above-mentioned future applications. To give an idea (though not exhaustive) of the state-of-the-art in the topic, the main results are presented here for ground and space observations and divided into intermediate-term and short-term precursors, respectively. A significant ground intermediate-term mechanical precursor is shown in Section 3.2 and a summary, even not exhaustive, of the principal characteristics of ground and space short-term SEME precursors is reported in sub-Section 3.3 (Tables 2 and 3).

Table 2: A summary of ground-based short-term (hour-days) SEME precursors

Observations	Modeling
EKE changes.	Streaming Potentials by saline water moving through porous rocks [62, 63].
B field changes.	
Ground potentials [59–61].	Stress applied effects to rocks containing piezoelectric materials [64–67].

ULF-ELF SEME [68–71]. ULF-HF SEME [26, 72–77]. VLF E field changes [78, 79].	EM behaviour of rocks [55, 80–83]. Rocks become a source of highly mobile electric carriers that increase electric conductivity and propagate through the rock as a charge cloud [84].
E field changes and gas emissions by rock microfracturing [31, 85, 86].	Number and dimensions of microcracks and redistribution of pore fluids. Motion of saline pore fluids and formation of intergranular water film [55, 86].
Low-frequency SEME Laboratory investigations about conversion of accumulated preseismic elastic energy to EM energy [87].	Rock as "igneous rock battery" due to the activation of positive hole charge carriers by stress. Dislocation movement leading to bond breaking of Si–OO–Si peroxy links [87].
Mechanical and EM signals from laboratory to geophysical scale [88].	First attempts to justify effects of applied stress on rocks [88].

Table 3: A summary of space-based short-term (hour-days) SEME precursors. Symbol ⇒means then

Observations	Modeling
Ionospheric E, B fields changes [8, 51, 54, 89–101].	SEME-waves generation Lithospheric lowpass filter on ULF-HF-waves ULF-ELF SEME-waves may reach the Earth surface and enter into near-Earth space [86, 98, 99, 102–109].
Ionospheric plasma temperature and density changes TEC. Decrease at the ionospheric F2 peak foF2 [110–112].	ULF-ELF SEME-waves-Ionospheric plasma interaction mechanisms [71, 103, 113, 114].

SEME-waves. Van Allen radiation belt particle precipitation. PBs (Particle Bursts) [8, 18, 48–50, 53, 115–118].	Alfven-wave radiation (from DC to some hundred Hz) propagates along the geomagnetic field lines Resonant wave-particle interaction at the radiation belt boundary with trapped electrons and protons from a few MeV to several tens of MeV Particle precipitation as a result of pitch angle diffusion [7, 8, 50, 51, 92, 119].
Variations in the atmospheric conductivity profiles [98, 99].	Fair weather currents [98]. Modification of spectral content of ELF-VLF radio noise during lightning discharges [99].
ULF SEME-waves and VLF SEME-waves from Satellite Intercosmos-24 [89].	ULF emissions of 0.2 nT penetrate through the ionosphere cyclotron interaction with protons of 0.5–5 MeV near the magnetic equatorial plane Proton distribution function becomes unstable for the Cherenkov VLF radiation of 0.1–20 kHz [119].
Amplitude and phase variations of radio-signal propagating in the earth-ionosphere wave guide). Disturbances in Omega and Loran VLF radio-waves propagation [120–122].	Abnormal ionisation in the lower ionosphere [121].

Short-term electric field strength attenuation of the Radio Monte Carlo (RMC) LF radio-signal [26].	Tropospheric radio defocusing mechanisms [26].
Atmospheric anomalies caused by VHF SEME-waves [123].	Significant enhancement of VHF EM-waves beyond line-of-sight [123].

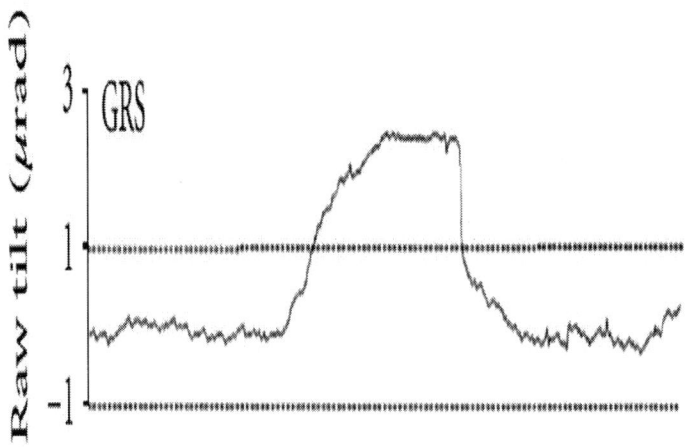

(a)

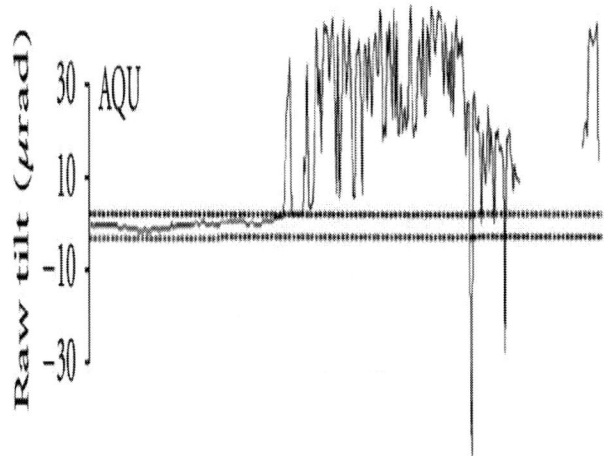

(b)

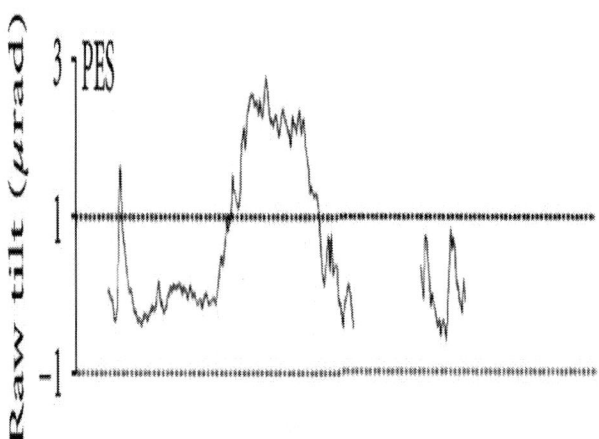

(c)

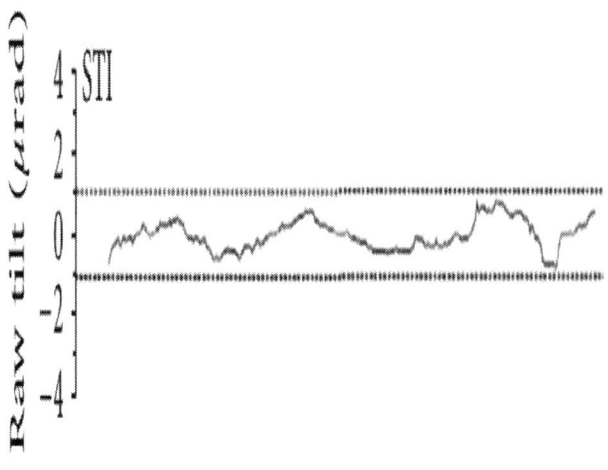

(d)

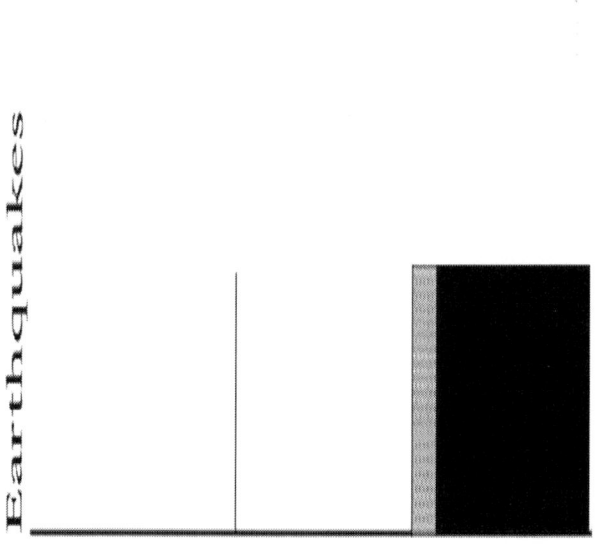

(e)

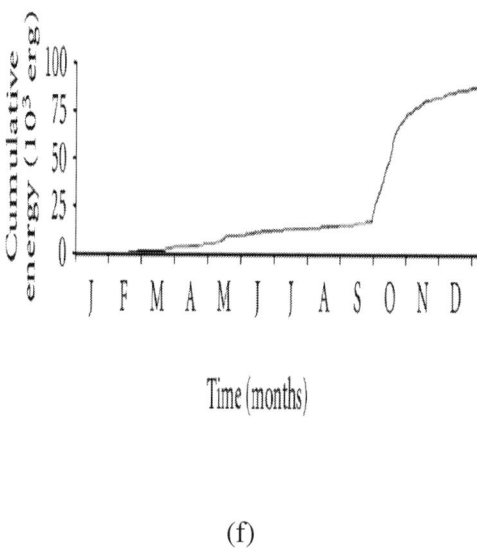

(f)

Figure 1: Original figure from the paper by Sgrigna and Malvezzi [16]. Fault creep events recorded during the year 1997 at the GRS (plot (a)), AQU (plot (b)), PES (plot (c)), STI (plot (d)) tilt sites, before the Sept 26, 1997 Umbria-Marches earthquakes (M=5.7; 6.0). Plot (e) shows selected earthquakes (e equal/greater than 10−8). A vertical bar marks a single event M4.4 occurred on May 12; two adjacent shadow and black rectangles of arbitrary amplitude represent two time intervals characterized by the occurrence of a preseismic swarm (lasting a few weeks with a peak event M4.4 on September 3), and of several thousands of aftershocks recorded in the following months, respectively. The two main shocks M5.7; 6.0 of September 26 occurred in the time interval between those marked by shadow and black areas. Plot (f) is the cumulative energy released by earthquakes in 1997.

Ground Creep-Related Intermediate-Term Precursors

A number of interesting results concerning anomalous surface tilt variations observed in local seismic regions during earthquake preparation have been reported over the years. They include the observation and modeling of creep-related tilt perturbations [16, 31, 32], precursory tilts detected before local and teleseismic earthquakes [29, 33], coseismic and postseismic tilts [34, 35]. These anomalies are easily detectable by tiltmeters [16, 31,36–38] and considered by many authors [17, 29, 31, 33, 39–43] to be intermediate-term earthquake precursors. The transmission of substantial stress over large distances has been debated [7, 16, 44].

Continuous hourly ground tilt data collected by the TELLUS tiltmeter network from 1981 to the present in the seismic region of the Central Apennines of Italy has systematically provided evidence of intermediate-term-earthquake tilt precursors [16]. An example of creep-related intermediate-term tilt precursors detected by the TELLUS network has been pointed out by Sgrigna and Malvezzi [16] on the occasion of the Umbria-Marches seismic sequence with two main shocks (M=5.7 and 6.0) with epicentres very close one each other (about 3 km) occurred on September 26, 1997, at 00:33 and 9:40 UTC, respectively. Figure 1 is the original figure taken from the paper by Sgrigna and Malvezzi [16] to which we invite to refer for the description of geodynamics of local crustal block system, characteristics of seismicity, and selection criteria for earthquakes and residual tilt signals.

The main features of the intermediate-term preseismic tilts reported in Figure 1 may be summarized as follows (Figure 1).

- Raw tilt data, filtered by meteorological and secular tectonic effects, revealed intermediate-term preseismic tilts with a shape, amplitude, and time duration similar to those already obtained in the same area [16, 31, 42, 43].

- Tilts are shifted in time relative to each other, indicating a possible propagation of the preseismic strain field from the preparation focal area to the tilt sites, through the rigid blocks of the region [26, 45] separated by inclined transition zones, filled by fault viscoelastic material [16, 29, 39, 46].

- A characteristic so-called site effect is evident in the signal amplification observed at the AQU tilt site when comparing amplitudes of this signal with those recorded at GRS and PES.

- Experimental values for the velocity of propagation are in agreement with previous results.

- The intermediate-term preseismic tilts have been interpreted as viscoelastic creep strains in the fault material, due to the propagation of stress-strain fields from the dilatant focal area to the observation sites.

- One-dimensional and two-dimensional numerical models have been proposed to justify qualitatively the main features (tilt anomaly shape and onset time delay and decay of anomaly amplitude with distance from the earthquake preparation zone) of the preseismic ground tilt behaviour [17, 26, 30]. Horizontal movements of rigid crustal blocks were also considered by Gabrielov et al. [47].

Ground and Space Short-Term Seismo-Associated EME Signals

Studies of seismoelectromagnetic emissions (SEME) have been developed for a few decades both at the Earth's surface and in the near-Earth space (atmosphere, ionosphere, and magnetosphere).

In recent years, interest has been increasing in the SEME signals consisting of a broad band (from approximately DC to a few tens of MHz) EM fields generated and transmitted by seismic sources into the near Earth's space before, during and after an earthquake. SEME characteristics and detectability as well as the effects they provoke in space (ionospheric and magnetospheric perturbations), have a very interesting and promising nature as a short-term earthquake predictor.

Several significant ground and space observations and modelling of such precursors are summarized in Tables 2 and 3, respectively.

Note that in the case of very shallow and strong earthquakes, when the size of the preparation focal zone is greater than the hypo-central depth (see relations (4) and (5)), also the higher frequency content of DC-HF SEME radiation could be transmitted from the Earth's surface to the near space.

Concerning radiation belt particle precipitation most preseismic PBs have been collected by satellites near the South Atlantic Anomaly (SAA) at altitudes generally between about 400 and 1200 km [48–50]. Moreover, the lower limit of the portion of the ionosphere-magnetosphere transition zone (i.e., the altitude where preseismic EME-waves may be captured in the geomagnetic field lines and, then, propagate up to the inner radiation belt) has been estimated from PBs space observations and resulted to be around 300–500 km [8, 51]. Besides, the lifetime of the longitudinal drift of PBs is determined by the particle loss rate during particle interaction with the residual atmosphere of the Earth. A lifetime of the order of several tens of minutes is obtained for electrons and protons of several tens of MeV [52]. During this time, particles may drift longitudinally around the Earth along the L-shell corresponding to the EME ground source location [50, 53].

This is a crucial factor for a possible use of preseismic PBs as an earthquake predictor since the longitudinal drift makes the PB detection possible by particle detectors installed on board satellites.

Another important factor is the opposite drift direction of positive- and negative-charged particles, which in principle could allow the location of EME wave-particle interaction zone (i.e., the PBs space source location) to be identified.

Nevertheless, there is still an open debate on the mechanism to be invoked in order to justify the phenomenology under study and, in particular, whether the very low amplitude ULF/ELF EM waves may reach the inner Van Allen radiation belt and cause the above-mentioned coupling phenomena. In fact, the electric and magnetic components of these EME-waves are estimated to be of only some fraction of mV/ m(Hz)1/2 and of some fraction of nT/(Hz)1/2 or less, respectively [54]. A qualitative representation of the space phenomenology is presented in Figure 2.

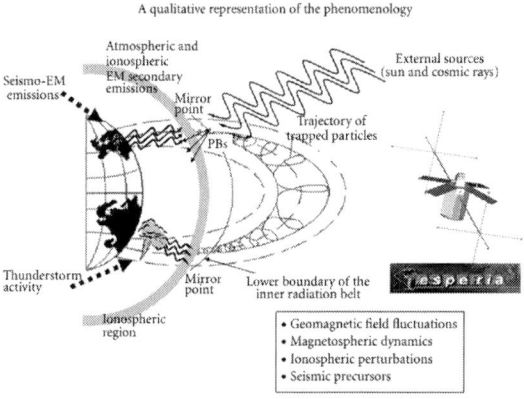

Figure 2: A schematic representation of the solid earth and near-Earth space (atmosphere, ionosphere, magnetosphere) with main associated physical phenomena: seismo-EM emissions, their propagation and interaction with

ionospheric plasma and magnetospheric trapped particles, cosmic rays and solar effects into the magnetosphere, and tropospheric TLE and TGF emissions. Trajectories of charged particles trapped by the geomagnetic field lines are represented in a meridian plane.

THE AUSONIA "SPACE SCIENTIFIC PLATFORM"

After a first satellite project named ESPERIA (Earthquake investigations by Satellite and Physics of the Environment Related to the Ionosphere and Atmosphere) was planned and designed for the Italian Space Agency (ASI) with objectives to only detect seismic precursors, a second more complete satellite project namedAUSONIA was proposed with aim at also studying other phenomena of the near-Earth space accompanying those associated with seismic events and which may interact with precursor signals. For a correct approach to an earthquake precursors study, all these signals must be recognized, isolated, and filtered from the data. A detailed technical description of the ESPERIA space mission concept can be found in the ASI Phase a Report [12] and in Sgrigna et al. 2008.

AUSONIA (Advanced mUlti-Instrument Satellite for a combined Observation of magNetosphere, Ionosphere, Atmosphere, and associated phenomena) is an Italian space project proposal submitted to the Italian Space Agency (ASI) within an ASI AO for earth observation [14]. AUSONIA was planned and designed by an Italian Consortium led by the Roma Tre University of Rome (Vittorio Sgrigna, Principal Investigator).

The aim of the AUSONIA project is to design and construct a small space platform planned with an multiinstrument payload and a LEO mini-satellite mainly concerned with the monitoring and mapping of the ionosphere-magnetosphere transition region. The scientific program is based on coordinated, continuous, and simultaneous space and ground-based observations, and on mutual data comparison with other missions of similar quality.

AUSONIA was proposed after the IUGG resolution in support of ESPERIA (2007 IUGG resolution N.5) (http://www.iugg.org/resolutions/), which welcomes the planning of several nations to launch ionospheric monitoring satellite missions. As mentioned above AUSONIA includes both the study of perturbative phenomena in the topside ionosphere (already planned for ESPERIA) and the field mapping of the same region to give a contribution in defining the IGRF and IRI models.

AUSONIA Scientific Aims

Scientific and methodological aspects of the AUSONIA space project are reported in Table 4.

Table 4: Science and methods of the Ausonia project

	Scientific objective	Expected results	International collaborations
Geomagnetic field mapping	Main field and secular variation will be the principal goals.	Contribution to the IGRF. A better knowledge of the Earth's core dynamics, secular variation, field inversions and crustal anomalies. 3D reconstruction of the mantle conductivity.	Synergy with SWARM mission, INGV ground network and SEGMA-ULF geomagnetic networks.
Monitoring of ionosphere and plasmasphere	Simultaneous measurements of local changes in the topside ionosphere and space and time variability of plasmasphere.	Contributions to the IRI model, ionospheric tomography, study of space weather events by in situ measurements and plasmaspheric TEC investigations.	Collaboration with NASA missions C/NOFS and STPSAT1. Use of CITRIS-like detector to collect signals from CERTO satellite and DORIS radio beacons terrestrial network. Comparisons with INGV and DIAS ionosonde data.

Detection of transient phenomena associated with thunderstorms	Detection of tropospheric transient luminous emissions (TLE), lightnings, terrestrial gamma ray flashes (TGF) and related energy transfer (~0,25–1 GW) from troposphere to iono-magnetosphere.	Understanding of TLE e TGF effects in the framework of the ionosphere-magnetosphere couplings.	Complementary observation campaigns of TLE e TGF phenomena to be carried out with the TARANIS satellite.
Study of iono-magnetospheric perturbations due to EM emissions of terrestrial origin	Study of the possible effects produced in the near-Earth space by EM emissions of seismic and volcanic origin.	The AUSONIA team can take profit from the expertise of the previous ESPERIA project (see the 2007 IUGG resolution N.5, http://www.iugg.org/resolutions.	The AUSONIA team is guest investigator of the DEMETER mission to study whistlers and radiation belt particles.
Investigation of Van Allen particle fluxes and tropospheric X/ rays	Study of temporal stability of the Van Allen radiation belts, detection of particle precipitation and tropospheric and cosmic X/ emissions.		A few key persons of the AGILE mission are also members of the AUSONIA team.

Experiments Planned on Board the AUSONIA Satellite

On the AUSONIA satellite are planned five experiments, MAGIA, ELECTRA, LUCE, CIELO, and TERRA. They are devoted to monitor geomagnetic field, plasma, and particle environment in the ionosphere-magnetosphere transition zone as well as to study optical/ UV and X/gamma emissions induced by tropospheric activity. The MAGIA (MAGnetic Instrument Array) experiment is constituted by a scalar, a fluxgate, and a 3-axes search-coil magnetometers to detect stationary, lower-frequency and higher-frequency magnetic field. The magnetometers are installed on the tips of two deployable booms (Boom_M_Right and Left, each one 5 meters long from the satellite spacecraft) to reduce the electromagnetic interference from the satellite equipments. The ELECTRA (ELECTRic field Analyser) experiment consists of 4 electric preamplified probes, each one installed on the tips of 4 meters deployable booms (4 meters long, called ELECTRA_Zenith, TAN, Right, Left) to allow to measure the 3 electric field components in the frequency range from about DC up to about 10 MHz). The MAGIA and ELECTRA sensors can highlight the correlation with lightnings and reconstruct the dynamics of the electromagnetic atmosphere-ionosphere. These measurements are also essential to study the LEP (lightning-induced electron precipitation) and all the phenomena of disturbance of the Van Allen belt-induced storms, in the AUSONIA project are included optical and UV detectors devoted to the observation of TLEs with high spatial and temporal resolution in specific frequency bands. Measurements are taken with video cameras and photometers with the hope of reconciling the need for high capture rate with the high-resolution image. The optical-UV for these observations are concentrated in the experiment LUCE in two separate blocks oriented to nadir and to limb, respectively. Each block consists of 2 cameras with filters optimized for the shooting of red sprites (VID1) and lightning (VID2), respectively, and 4 photometers (PH1, 2,3,4) for UV-visible measurements. The main characteristics of the LUCE experiment are summarized in Table 5. The precipitation of particles of the Van Allen belts was observed by several satellite missions, but many questions need an answer about the temporal and spatial stability of the Van Allen belts and the dynamics of interaction disturbances associated with magnetic storms, the electromagnetic emissions of tropospheric origin, the EM emissions of anthropogenic origin, and so forth. Other themes of topical scientific interest are the X and gamma emissions from the troposphere (TGF). They represent a background for satellite missions such as AGILE designed to explore gamma bursts from the sky. To study of these phenomena, the TERRA detector is designed to be installed on board the AUSONIA satellite. The experiment consists of two identical modules: TERRA_Nadir and TERRA_Tan oriented to Nadir and in the opposite the speed of the satellite, respectively. X- and gamma-ray detectors will be constantly active during the optical and EM measurements to allow to investigate the characteristics and origin of the TGF and their correlation with TLE. TERRA aims at revealing X- and gamma-ray bursts (TGF) from the Earth's troposphere. This to map TGF phenomena, to measure the X-ray spectrum observed range and determine the mechanisms that generate them, to observe the precipitation of particles from the Van Allen belts induced magnetic storms, tropospheric phenomena, seismoelectromagnetic emissions and emissions from anthropogenic EM, to measure range, direction, and temporal variation of the flow of precipitating charged particle, to reveal the runaway electrons, to study the interactions between whistler waves and trapped particles, to generate a trigger signal upon detection of a TGF and enable the acquisition of other experiments such as LUCE, to gather information on the length, height, changes in TGF, and to acquire a statistically significant amount of TGF events as a function of local time, geomagnetic conditions, and so forth.

Table 5: AUSONIA experiments and their positioning on board the satellite

Experiment		Module/Code	Probe/Instrument	Positioning/Pointing mode
Geomagnetic measurements	MAGIA (MACnetic Instrument array)	MAGIA_Scalar	Scalar magnetometer	Boom_M_Right
		MAGIA_Flux-Gate	Flux-gate Magnetometer	Boom_M_Right
		MAGIA_Search-Coil	Search-coil Magnetometer	Boom_M_Left
Electromagnetic measurements	ELECTRA (ELECTRic field analyser)	ELECTRA_Zenith	Electric probe	Boom_E_Zenit
		ELECTRA_Tan	Electric probe	Boom_E_Tan
		ELECTRA_DX	Electric probe	Boom_E_DX
		ELECTRA_SX	Electric probe	Boom_E_SX

Category	Experiment	Instrument	Type	Direction
Ultraviolet and Optical measurements	LUCE (transient LUminous emissions combined experiment)	LUCE_Nadir_VID1	Video-camera sprite	
		LUCE_Nadir_VID2	Video-camera lightning	Nadir
		LUCE_Nadir_PH1,2,3,4	Photometer 1,2,3,4	
		LUCE_Limb_VID1	Video-camera sprite	
		LUCE_Limb_VID2	Video-camera lightning	Limb
		LUCE_Limb_PH1,2,3,4	Photometer 1,2,3,4	
Plasma measurements	CIELO (Combined ionospheric experiment in low earth orbit)	CIELO_GPS	Gps	Zenith
		CIELO_PLASMA	LP, RPA, plasma driftmeter	Along the same direction of the satellite velocity
High-energy particles and X-gamma rays measurements	TERRA (circumTerrestrial high-energy paRticle and X-gamma ray analyser)	TERRA_Nadir	High-energy particles and X-gamma rays detector	Nadir
		TERRA_Tan	High-energy particles and X-gamma rays detector	In the opposite direction to the satellite velocity

Figure 3 illustrates the general satellite layout.

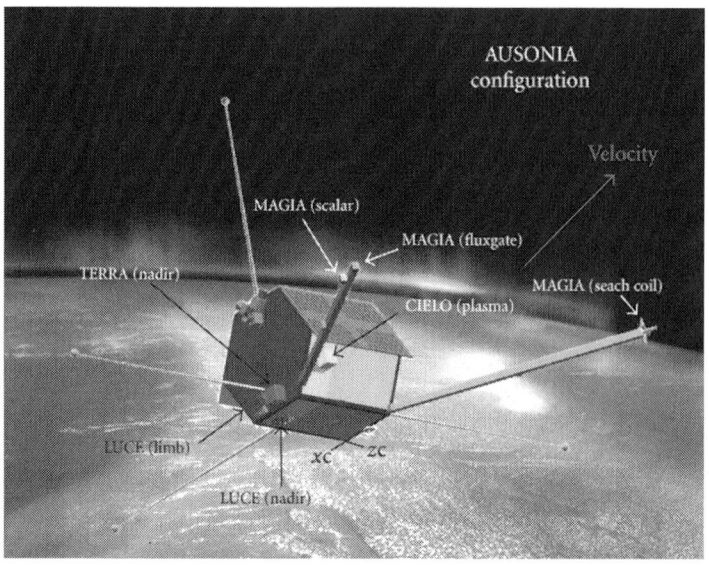

Figure 3: Schematic representation of the AUSONIA satellite project.

Planned experiments and instruments and their positioning on board the AURONIA satellite are reported in Table 5.

Mission Characteristics

At this overpreliminary step, the final parameters have not yet been completely defined. In Table 6, values are given for a MITA platform solution and a sun-synchronous orbit. The satellite orbit altitude has be chosen to optimise observations at the sunrise-sunset local time for a better identification of seismo-induced ionospheric disturbances. In fact, as reported by Molchanov and Hayakawa [55, 56] and Chuo et al. [57], an increase in the sporadic E-layer critical frequency at the terminator time (sunrise and sunset) is observed within 5 days before the earthquake that determines a corresponding increase in the D-layer electron density and a variation of the VLF propagation at the terminator time. Should AUSONIA be installed on board of another spacecraft, budgets, volume, orbit inclination, and altitude can be changed accordingly.

Table 6: AUSONIA satellite mission characteristics

Orbit	Sun-synchronous circular orbit 98° inclination
	Altitude between 600 to 800 km (TBD). See also notes after Table 3 in Section 3.3
	Revisit time: ≤24 h
Budgets	Power satellite total: ~270W (Payload total: 120W; platform total: 150W) (TBD)
	P/L data: ~306 kbps (36,5 Gbit of daily data, margin included) (TBD)
	Payload mass: ~120 kg (TBD)
Attitude orbit control system	Attitude determination: 0.001° (TBD)
	Attitude Accuracy (3 axes): 0.1° (TBD)
	3 reaction wheels, 3 magnetic coils, 2 star trackers, 3 gyroscopes, GPS receivers, three-axis magnetometer, 10 sun sensors
Spacecraft	Platform MITA or other platform of similar quality (TBD)
	Nadir pointing
	Thrusters applied to the platform (constant altitude and/or possible orbit changes) (TBC)
Mission duration	3 years

Comparisons between AUSONIA and Other Missions

In Table 7, the AUSONIA payload is compared with that of others missions of similar quality. It appears evident the AUSONIA capability in carrying out multiparametric measurements, as well as its character of "small scientific platform" for earth observation.

Table 7: AUSONIA instrument payload compared with that of other missions of similar quality

			AUSONIA	SWARM	TARANIS	DEMETER	FORMOSAT-2	ASIM (ISS)	VARIANT
Scalar magnetometer			X	X					
Flux-gate magnetometer			X	X					X
Search-coil magnetometer			X		X	X			X
Electric Probes			X	X	X	X			X
Langmuir probe			X	X		X			
Plasma driftmeter			X	X					
Optical-UV Detector	Nadir		X		X			X	
	Limb		X				X	X	
Particle Detectors	Nadir	Charged particles	X		(Low energy)				
		X/	X		X			(Low energy)	
	Limb	Charged particles	X			(Low energy)			
		X/	X					(Low energy)	

THE EGLE AND ARINA SPACE EXPERIMENTS

A few ESPERIA instruments (such as the particle detectors LAZIO and ARINA, and the search-coil magnetometer EGLE) have been built and tested in space [15, 19, 58]. EGLE was a technological demonstrator installed on board the International Space Station (ISS) on April 15, 2005, within the LAZIO-EGLE experiment of the ENEIDE mission, which has been coordinated by the European Space Agency (ESA) and received contributions from the Italian National Institute of Nuclear Physics (INFN) and Regione Lazio. The launch of ARINA occurred on June 15, 2006, within the PAMELA mission. ARINA will perform particle measurements on a quasipolar orbit RESURS DK-1 Russian LEO satellite. Data from ARINA, EGLE, and TELLUS may be studied together with those collected by DEMETER, through the Demeter Guest Investigator Programme.

The EGLE Magnetic Experiment on Board the International Space Station

The main goal of the EGLE experiment was to test in space an original very broad band search-coil magnetometer and associated data acquisition system based on the 1-Wire technology. The duration of the mission was of 10 days (15 April–25 April 2005).

The characteristics of the EGLE magnetometer are also important within the ISS applications. In fact, the monitoring of the EM environment on board the ISS needs both an appropriate observation methodology and a corresponding experimental equipment design. The continuous monitoring of the EM environment on board the ISS by an advanced magnetic experiment in the ULF-HF band is important in the following areas:

- search of space weather conditions in equatorial, middle-latitude, and subauroral ionosphere;
- geophysical research of plasma-wave processes connected to solar-magnetosphere-ionosphere-atmosphere-lithosphere interactions;

- investigation of the possible relationships between seismic activity and ULF-VLF phenomena possibly related to earthquakes;
- continuous monitoring of ULF-ELF-VLF activity in the near-Earth space including ELF-VLF pollution;
- Monitoring of natural and man-made variations of the plasma-sphere caused by whistlers.
- investigation of EM background and space weather phenomena;
- investigation of the effects of the large ISS structure on the propagating wave-front.

The LAZIO-EGLE experiment aims at performing measurements involving:

- the radiation environment;
- the magnetic environment inside the ISS.

The experiment includes the high-precision low-frequency magnetometer EGLE (Esperia's Geomagnetometer for a Low-frequency wave Experiment). EGLE is able to measure the intensity and variations in the magnetic field within the ISS and to correlate these measurements with those of particle fluxes. The study of these effects is important to detect electromagnetic field variations and particle pitch angle distribution of the precipitating particles. EGLE experiment is also the first test in space of a data acquisition system based on the 1-Wire technology.

The EGLE magnetometer consists of (Figure 4) the following

- a single axis search coil probe, the EGLE magnetometer head (MH);
- an electronic interface with amplifiers, filtering, and data acquisition unit (EGLE MB box);
- a 2-m long cable to connect LAZIO MEB and EGLE MB;(iv)a 1-Wire to RS232 serial adapter on the LAZIO pc tower.

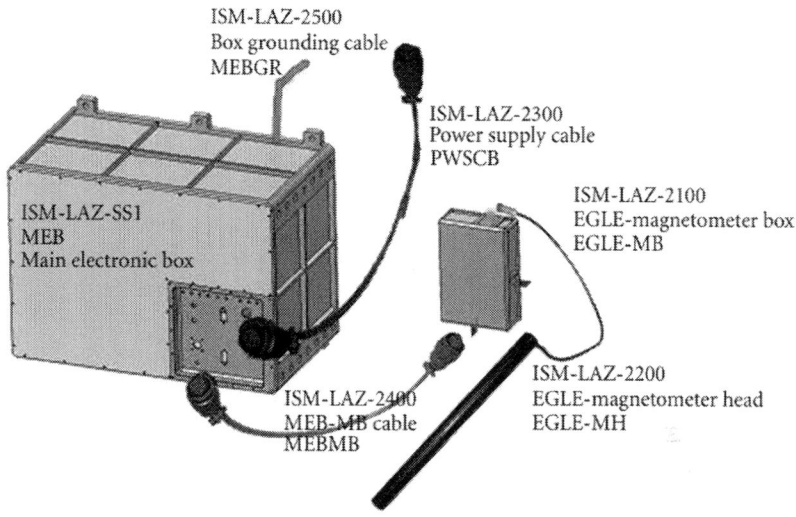

Figure 4: EGLE experimental setup.

Magnetic field signals detected by the EGLE-MH probe are amplified, filtered, and recorded by the EGLE acquisition and data handling board located in the EGLE-MB box. The EGLE magnetometer magnetic field data are collected in four frequency bands (DC through to 20 Hz raw data; 0.5–40 Hz; 500 Hz–5 kHz; 20–40 kHz integrated r.m.s. data).

Gaps between these frequency ranges have been chosen to filter well-known spurious artificial signals produced inside ISS.

The advantages of using EGLE device are:
- high-accuracy measurements;
- small dimensions and mass;
- low power consumption;
- data acquisition via 1-Wire technology;
- a standard ISS power supply of the device.

The peculiar characteristic of the 1-Wire technology prompted us to use it in the EGLE experiment to test its possible application in satellite EM measurements where the necessity to hold magnetic sensors far from the satellite body by expanding booms is an important factor for magnetic cleanness (see architecture of electric and magnetic probes in the ESPERIA payload). In fact, the use of 1-Wire technology can strongly reduce the numbers of wires necessary to connect many remote magnetic and electric probes (necessary in these types of investigations) with the central electronic unit located in the satellite body.

Figure 5 depicts the LAZIO-EGLE installation inside the PIRS section of the ISS. As can be seen, MEB (left), EGLE-MB (front), and EGLE-MH (right) are fixed by Velcro tags to the ISS wall. The characteristic frequency response of the EGLE probe is reported in Figure 6. An example of data recorded on board ISS is shown in Figure 7. As it can be seen, part of the ULF frequency band can also be detected by this sensor. This is an unusual characteristic for a search-coil probe and characterizes EGLE as an original broad-band magnetometer, which in a few satellite applications can allow a significant mass reduction by avoiding the use of flux-gate sensors.

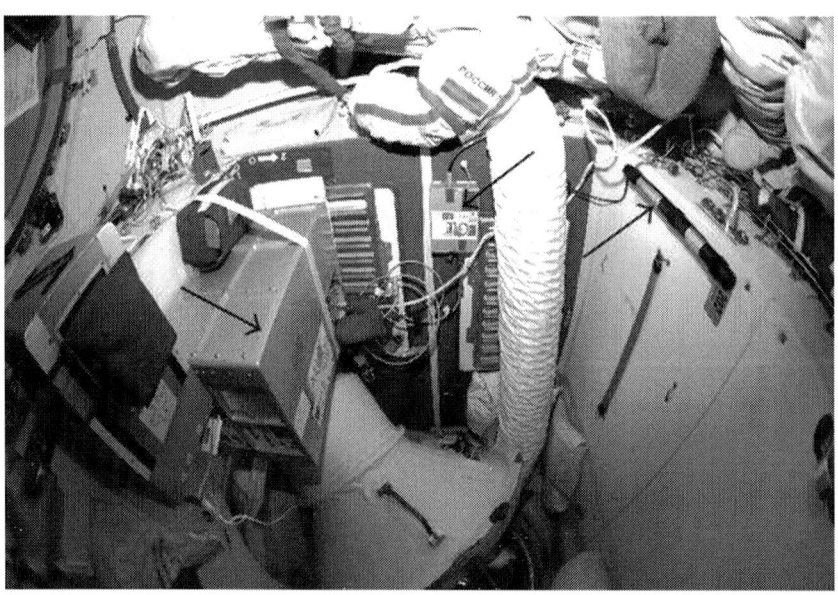

Figure 5: EGLE inside the PIRS module of the ISS. Arrows indicate MEB (left), EGLE-MB (front), and EGLE-MH (right).

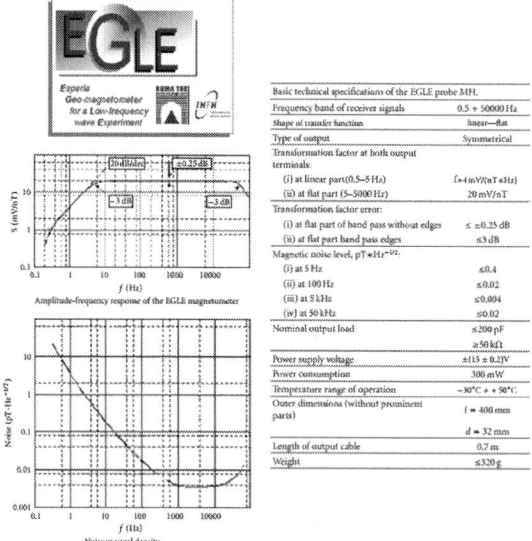

Figure 6: Frequency response and noise spectral density of the EGLE search-coil magnetometer together with technical specifications of the EGLE probe.

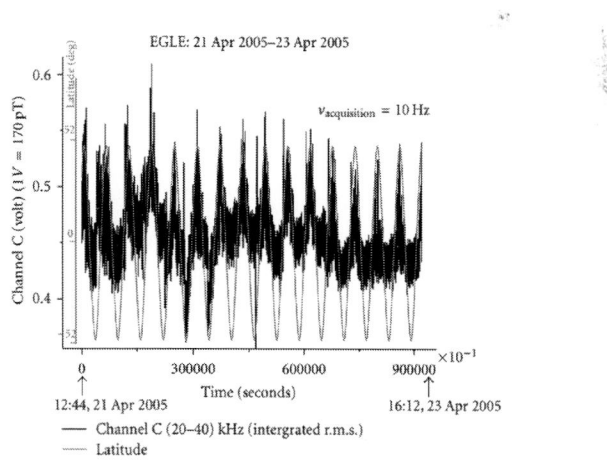

Figure 7: An example of magnetic data in the frequency band (20–40) kHz, recorded by the EGLE instrument on board the ISS during the period of the mission (15 April–25 April 2005). Superimposed to the signal is shown the latitude variation of the ISS.

The ARINA Particle Experiment on Board a LEO Satellite

The ARINA experiment consists of a proton-electron telescope to be installed on board the polar LEO Russian satellite RESURS-DK1 within the PAMELA mission. The orbit is elliptic, with an altitude ranging from 300 to 600 km and an inclination of 70.4 degree. The duration of the mission will be ≥3 years. The scientific objective of the experiment is to detect fluxes of high-energy charged particles (3 ÷ 100 MeV), from the inner radiation belt and to correlate them with seismic activity.

The main features of the ARINA instrument are reported in Figure 8. As can be seen from this figure, the instrument consists of a set of scintillation detectors C1–C12 made on the basis of polystyrene, which are viewed by photomultipliers (PMTs), the event recording system, the data acquisition and processing system (DAPS), the power supply system (PSS), and the command unit (CU). Detectors C1–C12 are functionally combined into three systems: the hodoscopic trigger system HTS (detectors C1–C3), the scintillation calorimeter SC (detectors C4–C9), and the anticoincidence system ACS (detectors C10–C12). Each of the detectors C1 and C2 consists of four strips directed perpendicularly and positioned just one under another. Detector C3 is situated below detectors C1 and C2 and has a mosaic structure (six elements). Each mosaic element is viewed by its own PMT. This type of assembly enables the angle of incident particle to be determined. The geometry and dimensions of detectors C1–C3 define the instrument aperture and the geometric factor. The scintillation calorimeter can comprise the detector C3 in addition to another set of detectors, C4–C9. It provides the separation of the protons and electrons and enables the particle energy to be measured by the number of detectors, passed by the particle up to its stop; that is, it is used the range of the particle in the stack of detectors. The ACS consists of the detector C10 and lateral detectors C11 and C12, and it is needed to exclude the particles moving in the opposite direction (from the bottom to upward) from being recorded as well as all directions beyond the aperture.

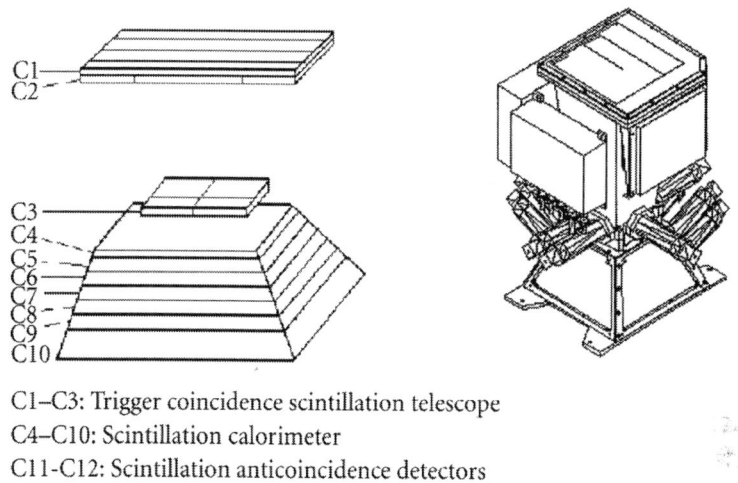

C1–C3: Trigger coincidence scintillation telescope
C4–C10: Scintillation calorimeter
C11-C12: Scintillation anticoincidence detectors

Figure 8: ARINA space instrument layout.

CONCLUSIONS

In this paper, we aimed at giving a contribution to earthquake precursor studies. At this purpose, ground and space observations and modeling have been presented together with specific space projects. In particular, we have clarified different methodological aspects on damage prevention and prediction approaches used to defend society from such destructive events as earthquakes and provided a short summary of the scientific background of ground and space observations on earthquake precursors together with relative first theoretical interpretations. Also a possible first empirical approach to deterministic earthquake prediction based on medium-term and short-term ground and space precursory phenomena has been given. The latter consists of EM emissions radiated from the Earth's surface and produced as a consequence of earthquake preparation and occurrence, or by human activities. They demonstrated to cause ionospheric perturbations that are detectable by LEO satellites. Within this framework, we have described the ESPERIA satellite project designed for detecting seismo-associated effects in the topside ionosphere and first ESPERIA instruments (LAZIO-EGLE and ARINA), which have been tested in space. But a field mapping of the topside ionosphere also demonstrated to be an important factor to contribute in defining both the IGRF and IRI magnetic and ionospheric models, as well as the monitoring of TLE and TGF tropospheric phenomena that have recently assumed a relevant importance. An IUGG resolution of 2007 in support of ESPERIA and, more generally, of an ionospheric mission with all the above elements as scientific objectives, triggered us in proposing the AUSONIA space project.

ACKNOWLEDGMENTS

The authors declare that they do not have any relation with all commercial devices mentioned in our paper, and that there is no conflict of interest for any of them.

REFERENCES

1. Japanese National Police Agency, Countermeasures for the Great East Japan Earthquake,http://www.npa.go.jp/archive/keibi/biki/ higaijokyo_e.pdf.

2. V. Sgrigna, A. Buzzi, L. Conti, P. Picozza, C. Stagni, and D. Zilpimiani, "Seismo-induced effects in the near-earth space: combined ground and space investigations as a contribution to earthquake prediction," Tectonophysics, vol. 431, no. 1–4, pp. 153–171, 2007. · ·

3. V. I. Keilis-Borok, "Intermediate-term earthquake prediction," Proceedings of the National Academy of Sciences of the United States of America, vol. 93, no. 9, pp. 3748–3755, 1996. · ·

4. V. I. Keilis Borok and A. Soloviev, Non Linear Dynamics of the Lithosphere and Earthquake Prediction, Springer, Berlin, Germany, 2003.

5. A. Peresan, V. Kossobokov, L. Romashkova, and G. F. Panza, "Intermediate-term middle-range earthquake predictions in Italy: a review," Earth-Science Reviews, vol. 69, no. 1-2, pp. 97–132, 2005. · ·

6. A. Tzanis and F. Vallianatos, "Distributed power-law seismicity changes and crustal deformation in the SW Hellenic ARC," Natural Hazards and Earth System Science, vol. 3, no. 3-4, pp. 179–195, 2003.

7. V. Sgrigna, R. Console, L. Conti, et al., "Preseismic natural emissions from the Earth's surface and their effects in the near earth space, A project for monitoring earthquakes from Space," American Geophysical Union, vol. 83, no. 19, article S356, 2002, abstract no. T22B-10.

8. V. Sgrigna, L. Carota, L. Conti et al., "Correlations between earthquakes and anomalous particle bursts from SAMPEX/PET satellite observations," Journal of Atmospheric and Solar-Terrestrial Physics, vol. 67, no. 15, pp. 1448–1462, 2005. ·

9. S. A. Pulinets, "Space technologies for short-term earthquake warning," Advances in Space Research, vol. 37, no. 4, pp. 643–652, 2006. · ·

10. M. Long, A. Lorenz, G. Rodgers, et al., "A cubesat derived design for a unique academic research mission in earthquake signature detection," in Proceedings of the 16th Annual/USU Conference on Small Satellites, pp. 1–17, Logan, Utah, USA, August 2002.

11. M. Parrot, "The micro-satellite DEMETER," Journal of Geodynamics, vol. 33, no. 4-5, pp. 535–541, 2002. · ·

12. V. Sgrigna, "(Principal Investigator), esperia science team, ESPERIA phase a report," Italian Space Agency (ASI), Program for Scientific Missions dedicated to Earth Sciences, Rome, Italy, 2001.

13. V. Sgrigna, A. Buzzi, L. Conti, P. Picozza, C. Stagni, and D. Zilpimiani, "The ESPERIA satellite project for detecting seismic-associated effects in the topside ionosphere. First instrumental tests in space,"Earth Planets and Space, vol. 60, pp. 463–475, 2009.

14. V. Sgrigna, the Ausonia Collaboration, The AUSONIA space project, Proposal submitted to the Italian Space Agency, 2008.

15. P. Picozza, (PAMELA/ARINA collaboration), The PAMELA Mission, 2003,http://wizard.roma2.infn.it/pamela/index.htm.

16. V. Sgrigna and V. Malvezzi, "Preseismic creep strains revealed by ground tilt measurements in central Italy on the occasion of the 1997 Umbria-Marche Apennines earthquake sequence," Pure

and Applied Geophysics, vol. 160, no. 8, pp. 1493–1515, 2003.
· ·

17.   V. Sgrigna, C. D'ambrosio, and T. B. Yanovskaya, "Numerical modeling of preseismic slow movements of crustal blocks caused by quasi-horizontal tectonic forces," Physics of the Earth and Planetary Interiors, vol. 129, pp. 313–324, 2002.

18.   L. Conti, A. Buzzi, and A. M. Galper, "Influence of the seismic activity on the inner Van Allen radiation belt," in Proceedings of the 10th Scientific Assembly of the International Association of Geomagnetism and Aeronomy (IAGA ‹05), p. 46, Toulose, France, July 2005, Session Division I, GA101: Monitoring earthquakes and volcanic activity by magnetic, electric and electromagnetic methods; IAGA2005-A-01518.

19.   V. Sgrigna, "Description and testing of ARINA and LAZIO/EGLE instruments in space within the ESPERIA mission project and the DEMETER guest investigation programme," in DEMETER Guest Investigator Workshop, Paris, France, May 2005.

20.   T. Lay and T. C. Wallace, Modern Global Seismology, Academic Press, San Diego, Calif, USA, 1995.

21.   A. M. Nur, "Dilatation, pore fluids and premonitory variation of TP/TS travel time," Bulletin of the Seismological Society of America, vol. 62, pp. 1217–1222, 1972.

22.   C. H. Scholz, "A physical interpretation of the Haicheng earthquake prediction," Nature, vol. 267, no. 5607, pp. 121–124, 1977. · ·

23.   V. I. Mjachkin, W. F. Brace, G. A. Sobolev, and J. H. Dieterich, "Two models for earthquake forerunners," Pure and Applied Geophysics PAGEOPH, vol. 113, no. 1, pp. 169–181, 1975. · ·

24.   I. P. Dobrovolsky, S. I. Zubkov, and V. I. Miachkin, "Estimation of the size of earthquake preparation zones," Pure and Applied Geophysics PAGEOPH, vol. 117, no. 5, pp. 1025–1044, 1979.
· ·

25.   I. P. Dobrovolsky, N. I. Gershenzon, and M. B. Gokhberg, "Theory of electrokinetic effects occurring at the final stage in the preparation of a tectonic earthquake," Physics of the Earth and Planetary Interiors, vol. 57, no. 1-2, pp. 144–156, 1989.

26.   F. Bella, M. Caputo, G. Della Monica et al., "Crustal blocks and seismicity in the Central Apennines of Italy," Nuovo Cimento

della Societa Italiana di Fisica C, vol. 21, no. 6, pp. 597–607, 1998.

27. T. Rikitake, "Earthquake precursors," Bulletin of the Seismological Society of America, vol. 65, pp. 1133–1162, 1975.

28. L. Conti, A. Cirella, V. Malvezzi, and V. Sgrigna, "A model for the propagation of preseismic electromagnetic fields through lithospheric and atmospheric media," in Proceedings of the 1st General Assembly, European Geosciences Union, p. 337, Nice, France, April 2004.

29. R. G. Bilham, "Delays in the onset times of near-surface strain and tilt precursor to earthquakes," inEarthquake Prediction: An International Review, P. J. Simpson and P. G. Richards, Eds., pp. 411–421, Geophysical Union, Washington, DC, USA, 1981.

30. F. Bella, P. F. Biagi, M. Caputo et al., "Very slow-moving crustal strain disturbances," Tectonophysics, vol. 179, no. 1-2, pp. 131–139, 1990.

31. F. Bella, P. F. Biagi, M. Caputo et al., "Possible creep-related tilt precursors obtained in the Central Apennines (Italy) and in the Southern Caucasus (Georgia)," Pure and Applied Geophysics PAGEOPH, vol. 144, no. 2, pp. 277–300, 1995. · ·

32. S. McHugh and M. J. S. Johnston, "A review of observations and dislocation modeling of some creep-related tilt perturbations from central California," in Terrestrial and Space Techniques in earthquake Prediction, A. Vogel, Ed., pp. 181–201, Vieweg and Sohn, Braunschweig, Germany, 1979.

33. R. G. Bilham, J. Beavan, K. Evans, and K. Hurst, "Crustal deformation metrology at lamont-doherty geological observatory," Earthquake Prediction Research, vol. 3, pp. 391–411, 1985.

34. W. Thatcher and N. Fujita, "Deformation of the mikata rhombus: strain buildup following the 1923 kanto earthquake, Central Honshu, Japan," Journal of Geophysical Research, vol. 89, pp. 2102–2106, 1984.

35. S. Ozawa, T. Nishimura, H. Suito, T. Kobayashi, M. Tobita, and T. Imakiire, "Coseismic and postseismic slip of the 2011 magnitude-9 Tohoku-Oki earthquake," Nature, vol. 475, no. 7356, pp. 373–377, 2011. · ·

36. C. E. Mortensen and M. J. S. Johnston, "The nature of surface tilt along 85 km of the San Andreas fault-preliminary results form a

14-instrument array," Pure and Applied Geophysics PAGEOPH, vol. 113, no. 1, pp. 237–249, 1975. · ·

37. R. G. Bilham and R. J. Beavan, "Strains and tilts on crustal blocks," Tectonophysics, vol. 52, no. 1–4, pp. 121–138, 1979.

38. A. Nur, H. Ron, and O. Scotti, "Fault mechanics and the kinematics of block rotations," Geology, vol. 14, no. 9, pp. 746–749, 1986.

39. Y. Ida, "Slow-moving deformation pulses along tectonic faults," Physics of the Earth and Planetary Interiors, vol. 9, no. 4, pp. 328–337, 1974.

40. A. K. Pevnev, "Earthquake prediction: geodetic aspects of the problem," Izvestija Akademija Nauk SSSR. Fizika Zemli, vol. 12, pp. 88–98, 1988.

41. A. K. Pevnev, "Deterministic geodetic prediction of preparation areas of strong crustal earthquakes,"Earthquake Prediction, vol. 11, pp. 11–23, 1989.

42. F. Bella, R. Bella, P. F. Biagi, A. Ermini, and V. Sgrigna, "Possible precursory tilts preceding some earthquakes ($3.0 \leq M \leq 3.8$) occurred in Central Italy between February 1981 and June 1983," Earthquake Prediction Research, vol. 4, pp. 147–154, 1986.

43. F. Bella, P. F. Biagi, M. Caputo, G. Della Monica, A. Ermini, and V. Sgrigna, "Ground Tilt anomalies accompanying the main earthquakes occurred in the central apennines (Italy) during the period 1986–1989," Il Nuovo Cimento C, vol. 16, no. 4, pp. 393–406, 1993. · ·

44. R. J. Geller, "Debate on VAN," Geophysical Research Letters, vol. 23, no. 11, 1996.

45. F. Salvini, "Block tectonics in thin-skin style-deformed regions: examples from structural data in the central apennines," Annali di Geofisica, vol. 36, pp. 97–109, 1993.

46. F. Bella, P. F. Biagi, A. Ermini, V. Sgrigna, and P. Manjgaladze, "Possible propagation of tilt and strain anomalies: velocity and other characteristics," Earthquake Prediction Research, vol. 4, pp. 195–209, 1986.

47. A. M. Gabrielov, T. A. Levshina, and I. M. Rotwain, "Block model of earthquake sequence," Physics of the Earth and Planetary Interiors, vol. 61, no. 1-2, pp. 18–28, 1990.

48. V. P. Pustovetov and A. B. Malyshev, "Space-time correlation of earthquakes and high-energy particle flux variations in the inner radiation belt," Cosmic Research, vol. 31, pp. 84–90, 1993.

49. E. A. Ginzburg, A. B. Malishev, I. P. Proshkina, and V. P. Pustovetov, "Correlation of strong earthquakes with radiation belt particle flux variations," Geomagn Aeronomy, vol. 34, pp. 315–320, 1994.

50. A. M. Galper, S. V. Koldashov, and S. A. Voronov, "High energy particle flux variations as earthquake predictors," Advances in Space Research, vol. 15, no. 11, pp. 131–134, 1995.

51. S. Y. Aleksandrin, A. M. Galper, L. A. Grishantzeva et al., "High-energy charged particle bursts in the near-Earth space as earthquake precursors," Annales Geophysicae, vol. 21, no. 2, pp. 597–602, 2003.

52. M. Walt, Introduction to Geomagnetically Trapped Radiation, Cambridge University Press, 1994.

53. M. E. Aleshina, S. A. Voronov, A. M. Galper, et al., "Correlation between earthquake epicenters and regions of high-energy particle precipitations from the radiation belt," Cosmic Research, vol. 30, no. 1, pp. 79–83, 1992.

54. M. Parrot, J. Achache, J. J. Berthelier et al., "High-frequency seismo-electromagnetic effects," Physics of the Earth and Planetary Interiors, vol. 77, no. 1-2, pp. 65–83, 1993.

55. O. A. Molchanov and M. Hayakawa, "On the generation mechanism of ULF seismogenic electromagnetic emissions," Physics of the Earth and Planetary Interiors, vol. 105, no. 3-4, pp. 201–210, 1998.

56. O. A. Molchanov and M. Hayakawa, "Subionospheric VLF signal perturbations possibly related to earthquakes," Journal of Geophysical Research, vol. 103, pp. 17489–17504, 1998.

57. Y. J. Chuo, J. Y. Liu, S. A. Pulinets, and Y. I. Chen, "The ionospheric perturbations prior to the Chi-Chi and Chia-Yi earthquakes," Journal of Geodynamics, vol. 33, no. 4-5, pp. 509–517, 2002.

58. V. Sgrigna, F. Altamura, S. Ascani et al., "First data from the EGLE experiment onboard the ISS," Microgravity Science and Technology, vol. 19, no. 5-6, pp. 70–74, 2007.

59. M. J. S. Johnston and R. J. Mueller, "Seismomagnetic observation during the 8 July 1986 magnitude 5.9 North Palm Springs earthquake," Science, vol. 237, no. 4819, pp. 1201–1203, 1987.

60. P. Varotsos, K. Alexopoulos, M. Lazaridou-Varotsou, and T. Nagao, "Earthquake predictions issued in Greece by seismic electric signals since February 6, 1990," Tectonophysics, vol. 224, no. 1–3, pp. 269–288, 1993.

61. K. Nomikos, F. Vallianatos, I. Kaliakatsos, E. Sideris, and M. Bakatsakis, "The latest aspects of telluric and electromagnetic variations associated with shallow and intermediate depth earthquakes in the South Aegean," Annali di Geofisica, vol. 40, no. 2, pp. 361–374, 1997.

62. A. B. Draganov, U. S. Inan, and Y. N. Taranenko, "ULF magnetic signatures at the Earth surface due to ground water flow: a possible precursor to earthquakes," Geophysical Research Letters, vol. 18, no. 6, pp. 1127–1130, 1991.

63. Y. Bernabé, "Streaming potential in heterogeneous networks," Journal of Geophysical Research B, vol. 103, no. 9, pp. 20827–20841, 1998.

64. J. R. Bishop, "Piezoelectric effects in quartz-rich rocks," Tectonophysics, vol. 77, no. 3-4, pp. 297–321, 1981.

65. P. Varotsos, N. Sarlis, M. Lazaridou, and P. Kapiris, "Transmission of stress induced electric signals in dielectric media," Journal of Applied Physics, vol. 83, no. 1, pp. 60–70, 1998.

66. F. Freund, "Charge generation and propagation in igneous rocks," Journal of Geodynamics, vol. 33, no. 4-5, pp. 543–570, 2002. ·
·

67. I. Stavrakas, C. Anastasiadis, D. Triantis, and F. Vallianatos, "Piezo stimulated currents in marble samples: precursory and concurrent-with-failure signals," Natural Hazards and Earth System Science, vol. 3, no. 3-4, pp. 243–247, 2003.

68. Y. A. Kopytenko, T. G. Matiashvili, P. M. Voronov, E. A. Kopytenko, and O. A. Molchanov, "Detection of ultra-low-frequency emissions connected with the Spitak earthquake and its aftershock activity, based on geomagnetic pulsations data at Dusheti and Vardzia observatories," Physics of the Earth and Planetary Interiors, vol. 77, no. 1-2, pp. 85–95, 1993.

69. A. C. Fraser-Smith, P. R. McGill, R. A. Helliwell, and O. G. Villard, "Ultra low frequency magnetic field measurements in southern California during the Northridge Earthquake of 17 January 1994,"Geophysical Research Letters, vol. 21, no. 20, pp. 2195–2198, 1994.

70. V. S. Ismaguilov, Y. A. Kopytenko, K. Hattori, P. M. Voronov, O. A. Molchanov, and M. Hayakawa, "ULF magnetic emissions connected with under sea bottom earthquakes," Natural Hazards and Earth System Sciences, vol. 1, pp. 23–31, 2001.

71. K. Ohta, K. Umeda, N. Watanabe, and M. Hayakawa, "ULF/ELF emissions observed in Japan, possibly associated with the Chi-Chi earthquake in Taiwan," Natural Hazards and Earth System Sciences, vol. 1, pp. 37–42, 2001.

72. J. W. Warwick, C. Stoker, and T. R. Meyer, "Radio emission associated with rock fracture: possible application to the great Chilean Earthquake of May 22, 1960," Journal of Geophysical Research, vol. 87, no. 4, pp. 2851–2859, 1982.

73. K. Oike and T. Ogawa, "Electromagnetic radiations from shallow earthquakes observed in the LF range,"Journal of Geomagnetism & Geoelectricity, vol. 38, no. 10, pp. 1031–1040, 1986.

74. M. J. S. Johnston, "Review of electric and magnetic fields accompanying seismic and volcanic activity,"Surveys in Geophysics, vol. 18, no. 5, pp. 441–475, 1997.

75. S. Uyeda, K. S. Al-Damegh, E. Dologlou, and T. Nagao, "Some relationship between VAN seismic electric signals (SES) and earthquake parameters," Tectonophysics, vol. 304, no. 1-2, pp. 41–55, 1999. · ·

76. K. Eftaxias, P. Kapiris, J. Polygiannakis et al., "Experience of short term earthquake precursors with VLF-VHF electromagnetic emissions," Natural Hazards and Earth System Science, vol. 3, no. 3-4, pp. 217–228, 2003.

77. F. Vallianatos, D. Triantis, A. Tzanis, C. Anastasiadis, and I. Stavrakas, "Electric earthquake precursors: from laboratory results to field observations," Physics and Chemistry of the Earth, vol. 29, no. 4–9, pp. 339–351, 2004. · ·

78. A. Nardi and M. Caputo, "Perspective electric earthquake precursors observed in the Apennines," inProceedings of the 8th

Workshop on Non Linear Dynamics and Earthquake Prediction, ICTP, October 2005.

79.   A. Nardi and M. Caputo, "Perspective electric earthquake precursors observed in the Apennines,"Bollettino di Geofisica Teorica ed Applicata, vol. 47, pp. 3–12, 2006.

80.   S. K. Park, M. J. S. Johnston, T. R. Madden, F. D. Morgan, and H. F. Morrison, "Electromagnetic precursors to earthquakes in the ulf band: a review of observations and mechanisms," Reviews of Geophysics, vol. 31, no. 2, pp. 117–132, 1993.

81.   M. Merzer and S. L. Klemperer, "Modeling low-frequency magnetic-field precursors to the Loma Prieta earthquake with a precursory increase in fault-zone conductivity," Pure and Applied Geophysics, vol. 150, no. 2, pp. 217–248, 1997.

82.   V. Surkov, "ULF electromagnetic perturbations resulting from the fracture and dilatancy in the earthquake preparation zone," in Atmospheric and Ionospheric Electromagnetic Phenomena Associated with Earthquakes, M. Hayakawa, Ed., pp. 371–382, TERRAPUB, Tokyo, Japan, 1999.

83.   M. Hayakawa, Y. Kopytenko, N. Smirnova, V. Troyan, and T. Peterson, "Monitoring ULF magnetic disturbances and schemes for recognizing earthquake precursors," Physics and Chemistry of the Earth, Part A, vol. 25, no. 3, pp. 263–269, 2000. · ·

84.   F. Freund, "On the electrical conductivity structure of the stable continental crust," Journal of Geodynamics, vol. 35, no. 3, pp. 353–388, 2003. · ·

85.   G. Areshidze, F. Bella, P. F. Biagi et al., "Anomalies in geophysical and geochemical parameters revealed on the occasion of the Paravani (M=5.6) and Spitak (M=6.9) earthquakes (Caucasus)," Tectonophysics, vol. 202, no. 1, pp. 23–41, 1992.

86.   Z. Guo, B. Liu, and Y. Wang, "Mechanism of electromagnetic emissions associated with microscopic and macroscopic cracking in rocks," in Electromagnetic Phenomena Related to Earthquake Prediction, M. Hayakawa, Ed., pp. 523–529, TERRAPUB, Tokyo, Japan, 1994.

87.   F. T. Freund, A. Takeuchi, and B. W. S. Lau, "Electric currents streaming out of stressed igneous rocks—a step towards understanding pre-earthquake low frequency EM emissions,"

Physics and Chemistry of the Earth, vol. 31, no. 4–9, pp. 389–396, 2006. · ·

88. K. Eftaxias, V. Sgrigna, and T. Chelidze, "Mechanical and electromagnetic phenomena accompanying pre-seismic deformation: from laboratory to geophysical scale," Tectonophysics, vol. 431, no. 1–4, pp. 1–5, 2007. · ·

89. O. A. Molchanov, O. A. Mazhaeva, A. N. Golyavin, and M. Hayakawa, "Observation by the Intercosmos-24 satellite of ELF-VLF electromagnetic emissions associated with earthquakes," Annales Geophysicae, vol. 11, pp. 431–440, 1993.

90. C. J. Rodger, R. L. Dowden, and N. R. Thomson, "Observations of electromagnetic activity associated with earthquakes by low-altitude satellites," in Atmospheric and Ionospheric Electromagnetic Phenomena Associated with Earthquakes, M. Hayakawa, Ed., pp. 697–710, TERRAPUB, Tokyo, Japan, 1999.

91. M. B. Gokhberg, V. A. Morgounov, and E. L. Aronov, "On the high frequency electromagnetic radiation during seismic activity," Dokladi Akademii Nauk USSR, vol. 248, pp. 1077–1081, 1979.

92. V. I. Larkina, V. V. Migulin, O. A. Molchanov, I. P. Kharkov, A. S. Inchin, and V. V. Schvetsova, "Some statistical results on very low frequency radiowave emissions in the upper ionosphere over earthquake zones," Physics of the Earth and Planetary Interiors, vol. 57, pp. 100–109, 1989.

93. M. Parrot and M. M. Mogilevsky, "VLF emissions associated with earthquakes and observed in the ionosphere and the magnetosphere," Physics of the Earth and Planetary Interiors, vol. 57, no. 1-2, pp. 86–99, 1989.

94. S. V. Bilichenko, F. S. Iljin, E. F. Kim, et al., "ULF response of the ionosphere for earthquake preparation processes," Doklady Akademii Nauk USSR, vol. 311, pp. 1077–1080, 1990.

95. O. N. Serebryakova, S. V. Bilichenko, V. M. Chmyrev, et al., "Electromagnetic ELF radiation from earthquake regions as observed by low-altitude satellite," Geophysical Research Letters, vol. 19, pp. 91–94, 1992.

96. V. M. Chmyrev, N. V. Isaev, O. N. Serebryakova, V. M. Sorokin, and Y. P. Sobolev, "Small-scale plasma inhomogeneities and

correlated ELF emissions in the ionosphere over an earthquake region," Journal of Atmospheric and Solar-Terrestrial Physics, vol. 59, no. 9, pp. 967–974, 1997.

97. C. C. Lee, J. Y. Liu, C. J. Pan, and K. Igarashi, "The heights of sporadic-E layer simultaneously observed by the VHF radar and ionosondes in Chung-Li," Geophysical Research Letters, vol. 27, no. 5, pp. 641–644, 2000. · ·

98. S. A. Pulinets, K. A. Boyarchuk, V. V. Hegai, V. P. Kim, and A. M. Lomonosov, "Quasielectrostatic model of atmosphere-thermosphere-ionosphere coupling," Advances in Space Research, vol. 26, no. 8, pp. 1209–1218, 2000. · ·

99. M. Hayakawa, O. A. Molchanov, and A. P. Nikolaenko, "Model variations in atmospheric radio noise caused by preseismic modifications of tropospheric conductivity profile," in Seismo Electromagnetics: Lithosphere-Atmosphere-Ionosphere Coupling, M. Hayakawa and O. A. Molchanov, Eds., pp. 349–352, TERRAPUB, Tokyo, Japan, 2002.

100. A. Buzzi, L. Conti, A. M. Galper, et al., "Sismo-electromagnetic emissions," in Proceedings of the NATO Advances Study. Institute on 'Sprites, Elves and Intense Lightning Discharges', M. Fullekrug, E. A. Mareev, and M. J. Rycroft, Eds., vol. 225 of NATO Science Series II: Mathematics, Physics and Chemistry, pp. 388–389, Springer, 2006.

101. M. Parrot, "Statistical study of ELF/VLF emissions recorded by a low-altitude satellite during seismic events," Journal of Geophysical Research, vol. 99, pp. 23339–23347, 1994.

102. M. A. Fenoglio, M. J. S. Johnston, and J. D. Byerlee, "Magnetic and electric fields associated with changes in high pore pressure in fault zones: application to the Loma Prieta ULF emissions," Journal of Geophysical Research, vol. 100, no. 7, pp. 12–958, 1995.

103. O. A. Molchanov, M. Hayakawa, and V. A. Rafalsky, "Penetration characteristics of electromagnetic emissions from an underground seismic source into the atmosphere, ionosphere and magnetosphere,"Journal of Geophysical Research, vol. 100, pp. 1691–1712, 1995.

104. R. Teisseyre, "Generation of electric field in an earthquake preparation zone," Annali di Geofisica, vol. 40, no. 2, pp. 297–304, 1997.

105. V. V. Grimalsky, I. A. Kremenetsky, and Y. G. Rapoport, "Excitation of EMW in the lithosphere and propagation into magnetosphere," in Atmospheric and Ionospheric Electromagnetic Phenomena Associated with Earthquakes, M. Hayakawa, Ed., pp. 777–787, TERRAPUB, Tokyo, Japan, 1999.

106. F. Vallianatos and A. Tzanis, "A model for the generation of precursory electric and magnetic fields associated with the deformation rate of the earthquake focus," in Atmospheric and Ionospheric Electromagnetic Phenomena Associated with Earthquakes, M. Hayakawa, Ed., pp. 287–305, TERRAPUB, Tokyo, Japan, 1999.

107. V. M. Sorokin, V. M. Chmyrev, and A. K. Yaschenko, "Electrodynamic model of the lower atmosphere and the ionosphere coupling," Journal of Atmospheric and Solar-Terrestrial Physics, vol. 63, pp. 1681–1691, 2001.

108. N. Gershenzon and G. Bambakidis, "Modeling of seismo-electromagnetic phenomena," Russian Journal of Earth Sciences, vol. 3, pp. 247–275, 2001.

109. Y. Fujinawa, T. Matsumoto, and K. Takahashi, "Modeling confined pressure changes inducing anomalous electromagnetic fields related with earthquakes," Journal of Applied Geophysics, vol. 49, no. 1-2, pp. 101–110, 2002. · ·

110. J. Y. Liu, Y. I. Chen, Y. J. Chuo, and C. S. Chen, "A statistical investigation of preearthquake ionospheric anomaly," Journal of Geophysical Research A, vol. 111, no. 5, Article ID A05304, 2006. · ·

111. K. Heki, "Ionospheric electron enhancement preceding the 2011 Tohoku-Oki earthquake," Geophysical Research Letters, vol. 8, Article ID L17312, 5 pages, 2011. ·

112. H. Tsuji, Y. Hatanaka, T. Sagiya, and M. Hashimoto, "Coseismic crustal deformation from the 1994 Hokkaido-Toho-Oki earthquake monitored by a nationwide continuous GPS array in Japan," Geophysical Research Letters, vol. 22, no. 13, pp. 1669–1672, 1995. · ·

113. E. Blanc, "Observations in the upper atmosphere of infrasonic waves from natural or artificial sources: a summary," Annales Geophysicae, vol. 3, no. 6, pp. 673–688, 1985.

114. Y. Zaslavski, M. Parrot, and E. Blanc, "Analysis of TEC measurements above active seismic regions,"Physics of the Earth and Planetary Interiors, vol. 105, pp. 219–228, 1998.

115. A. M. Galper, V. B. Dimitrenko, N. V. Nikitina, V. M. Grachev, and S. E. Ulin, "Interrelation between high-energy charged particle fluxes in the radiation belt and seismicity of the earth," Cosmic Research, vol. 27, article 789, 1989.

116. S. A. Voronov, A. M. Galper, S. V. Koldashov, et al., "Increases in high energy charged particle fluxes near the South Atlantic magnetic anomaly and the seismicity of the earth," Cosmic Research, vol. 28, pp. 789–791, 1990.

117. Y. I. Galperin, V. A. Gladyshev, N. V. Jordjio, and V. I. Larkina, "Precipitation of high-energy captured particles in the magnetosphere above the epicenter of an incipient earthquake," Cosmic Research, vol. 30, pp. 89–106, 1992.

118. A. Buzzi, M. Parrot, and J. A. Sauvaud, "Precipitation of particles by intense electromagnetic harmonic waves during magnetic storms," in Proceedings of the International Demeter Workshop, Toulouse, France, June 2006.

119. V. V. Krechetov, "Cerenkov radiation of protons in the magnetosphere as a source of VLF waves preceding an earthquake," Geomagnetism and Aeronomy, vol. 35, no. 5, pp. 688–691, 1996.

120. M. Hayakawa and H. Sato, "Ionospheric perturbations associated with earthquakes, as detected by sub-ionospheric VLF propagation," in Electromagnetic Phenomena Related to Earthquake Prediction, M. Hayakawa and Y. Fujinawa, Eds., pp. 391–397, TERRAPUB, Tokyo, Japan, 1994.

121. V. A. Morgounov, T. Ondoh, and S. Nagai, "Anomalous variation of VLF signals associated with strong earthquakes M≥7.0," in Electromagnetic Phenomena Related to Earthquake Prediction, M. Hayakawa and Y. Fujinawa, Eds., pp. 409–428, TERRAPUB, Tokyo, Japan, 1994.

122. I. Gufeld, G. Gusev, and O. Pokhotelov, "Is the prediction of earthquake dates possible by the VLF radiowave monitoring method?" in Electromagnetic Phenomena Related to Earthquake Prediction, M. Hayakawa and Y. Fujinawa, Eds., pp. 381–389, TERRAPUB, Tokyo, Japan, 1994.

123. H. Fujiwara, M. Kamogawa, M. Ikeda et al., "Atmospheric anomalies observed during earthquake occurrences," Geophysical Research Letters, vol. 31, Article ID L17110, 4 pages, 2004.

Chapter 5

Electromagnetic Radiation Energy Harvesting – The Rectenna Based Approach

Gabriel Abadal[1], Javier Alda[2], and Jordi Agustí[1]

[1]Departament d'Enginyeria Electrònica, Escola d'Enginyeria, Universitat Autònoma de Barcelona, Barcelona, Spain
[2]Applied Optics Complutense Group Facultad de Óptica y Optometría, Universidad Complutense de Madrid, Madrid, Spain

INTRODUCTION

The Energy Available in the Electromagnetic Spectrum

How much energy is available around us? Which use can we give to this energy? These are two questions to which answers had been changing over time. What would be our particular answer if a forefather or an ancestor would ask them to us? Some sources of energy like sun, wind or sea waves have been present unaltered since the prehistoric times and before to nowadays. Some others like oil and natural gas have been progressively reduced by the action of man. But it is interesting to notice that there are some other sources, which we can name as artificial sources, and that have emerged by man's action, as a consequence of industrial and technological development. Such modern or artificial energy sources are directly connected to the energy harvesting technology since, for instance, most of the vibrations or temperature gradients are produced by machines and engines. Also in the electromagnetic spectrum, we can harvest energy not only from natural sun radiation, but also from all the artificial radiofrequency sources that are permanently increasing in number and which are a consequence of one of the last technological revolutions: the Information and Communications Technology (ICT) revolution.

Although when we think about electromagnetic (EM) waves at present time, we probably tend to think about examples like radio, TV or cell phones, where the information part of the electromagnetic signal is the protagonist, we should bear in mind that those signals are in fact a combination of information and energy. In this chapter, we are not interested in describing how information can be transmitted through electromagnetic waves but how the energy of these waves is transmitted and collected to be harvested and used to supply ICT devices. In order to calculate how much energy can be associated to an electromagnetic wave, we have to consider the physical nature of these particular waves.

Basic Concepts

Electromagnetic waves in the electromagnetic spectrum (figure 1) are characterized by their wavelength λ or, alternatively, by their frequency v. Both magnitudes are related with the propagation speed of such waves, the speed of light c, through:

$$c = \lambda \cdot v$$

(1)

On the other hand, the frequency of EM radiation is directly related to the energy E of a photon associated to this radiation, i.e. the quantum of EM radiation or the most fundamental constitutive part of this radiation as defined by quantum mechanics, by

$$E = h \cdot v$$

(2)

Where $h = 6.626 \ 10^{-34}$ J s is the Planck constant.

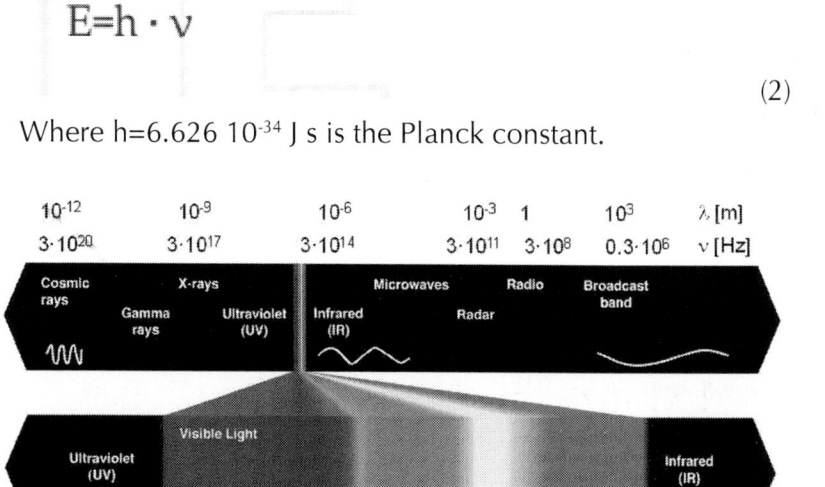

Figure 1: Diagram of the electromagnetic spectrum with indications of the wavelength, λ, and frequency, v, of the most representative radiations from shorter and most energetic, cosmic rays, to the longer and less energetic radiofrequencies. A zoom detail of the optical part of the spectrum shows that light radiations is in the hundred nm and THz range of wavelengths and frequencies respectively.

Unlike what occurs in photovoltaics technology, where optical radiation energy is better accounted in terms of photon energy since there the conversion mechanism is based in photon-electron interactions, in rectenna technology it is more convenient to express the input EM radiation in terms of the power or the power density of the EM wave.

An EM wave can be defined as a form of energy radiated by a source which results in a combination of oscillating electric and magnetic fields. In most of materials, the direction of the EM wave propagation is perpendicular to the electric and magnetic fields, which are also oscillating in phase perpendicular to each other.

The set of equations which describe how electric and magnetic fields propagate, interact and how they are influenced by material properties are Maxwell's equations. An EM wave can be described with these equations, which must be met for a set of particular boundary conditions. Maxwell's equations are summarized in Table 1.

Table 1: Maxwell's equations

Law	Integral form	Differential form
Faraday's law of induction	$\oint_c \vec{E} \cdot \vec{dl} = -\dfrac{\partial}{\partial t} \iint_s \vec{B} \cdot \vec{ds}$	$\nabla \times \vec{E} = -\dfrac{\partial}{\partial t}\vec{B}$
Ampère's circuital law	$\oint_c \vec{H} \cdot \vec{dl} = \iint_s \vec{J} \cdot \vec{ds} + \dfrac{\partial}{\partial t} \iint_s \vec{D} \cdot \vec{ds}$	$\nabla \times \vec{H} = \vec{J} + \dfrac{\partial}{\partial t}\vec{D}$
Gauss's law	$\oiint_s \vec{D} \cdot \vec{ds} = \iiint_v \rho \cdot dV$	$\nabla \times \vec{D} = \rho$
Gauss's law for magnetism	$\oiint_s \vec{B} \cdot \vec{ds} = 0$	$\nabla \times \vec{B} = 0$

Where $\vec{E}$ is the electric field intensity, $\vec{B}$ and $\vec{H}$ are the magnetic fields; $\vec{J}$ is the total current density,

$\vec{D}$ is the electric displacement field and ρ is the total charge density.

The propagation of a plane EM wave can be described by the EM wave equation, which can be derived from Maxwell's equations. The homogeneous form of this second-order differential equation can be written in terms of either the electric field or the magnetic field as

$$\left(\nabla^2 - \mu\varepsilon \frac{\partial^2}{\partial t^2} \right) \left\{ \begin{array}{c} E_y(x,t) \\ B_z(x,t) \end{array} \right\} = 0$$

(3)

Where μ and ε are the permeability and the permittivity of the propagation medium, respectively.

Knowing that EM waves carry energy with them in the form of electric and magnetic fields, we can compute their energy flow per unit area using the so called Poynting vector

$$\vec{S} = \vec{E} \times \vec{H}$$

(4)

From the Poynting vector and considering a uniform plane wave the time-average power density of the EM wave can be computed as [1]

$$P_{av} = \frac{1}{2} \cdot \frac{|E_0|^2}{\mathrm{Re}\{\eta\}}$$

(5)

Where E_0 is the peak value of the electric field and η is the impedance of the propagating medium. If the wave propagates in a loss-less dielectric medium η is a real number. Being this medium the free space, the impedance can be computed as follows:

$$\eta_0 = \sqrt{\frac{\varepsilon_0}{\mu_0}} = \frac{1}{\varepsilon_0 \cdot c_0}$$

(6)

Where μ_0 is the vacuum permeability, ε_0 is the vacuum permittivity and c0 is the speed of light in free space. The value for the impedance of the vacuum is about 377Ω.

A good approximation to the radiated power at a certain distance d from an emitter can be computed considering that the emitter is an isotropic radiator (EM point source which radiates the same power in all directions)

$$P_{rd} = \frac{P_{rT}}{4 \cdot \pi \cdot d^2}$$

(7)

Where P_{rT} is the total radiated power and d the distance from the emitter Notice that real antennas do not radiate isotropically, they have a certain radiation pattern which depends mainly on the geometry of the antenna and the surrounding media.

In 1999 the Council of the European Union made some recommendations on the limitation of exposure to electromagnetic fields [2]. Table 2 summarizes the maximum recommended values for the electric field.

Table 2: Reference levels for electric fields from 0 Hz to 300 GHz

Frequency range	E-field strength (V/m)
0-1 Hz	–
1-8 Hz	10000
8-25 Hz	10000
0.025-0.8 kHz	250 / f
0.8-3 kHz	250 / f
3-150 kHz	87
0.15-1 MHz	87
1-10 MHz	$87 / f^{0,5}$
10-400 MHz	28
400-2000 MHz	$1,375 \cdot f^{0,5}$
2-300 GHz	61

On the other hand, the IEEE International Committee on Electromagnetic Safety has made some additional recommendations in order to protect human beings from harmful effects caused by the exposure to electromagnetic fields [3]. Table 3 summarizes the maximum recommended values for the RMS electric field, magnetic field and power density.

Table 3: Reference levels for electric, magnetic fields from 0 Hz to 300 GHz

Frequency range (MHz)	RMS electric field strength (V/m)	RMS magnetic field strength (A/m)	RMS power density (E-field, H-field) (W/m²)
0.1-1	1842	16,3 / f	$(9000 , 100000 / f^2)$
1-30	1842 / f	16,3 / f	$(9000 / f^2 , 100000 / f^2)$

30-100	61,4	16,3 / f	$(10 , 100000 / f^2)$
100-300	61,4	0,163	10
300-3000	–	–	f / 30
3000-30000	–	–	100
30000-300000	–	–	100

Finally, although tables 2 and 3 give a good idea of the maximum energy available from RF emissions in terms of electric field and power density, in table 4 power density values and ranges corresponding to different applications are also summarized and compared to sunlight in the visible range.

Table 4: Comparison of power densities for different applications with solar radiation in the visible range

Application	Power density (mW/cm²)
Old UHF TV band	10^{-9}
FM radio @ 50 km from 100kW base station	10^{-7}
ISM bands: Zigbee/Bluetooth/WIFi	$10^{-8}/10^{-7}/10^{-6}$
Standard ambient level with no high power equipment	$10^{-6} – 10^{-5}$
GSM, UMTS (3G telecom) @ 10 m from base station	$10^{-6} – 10^{-4}$
Cellular phone @ 50 m from base station	$10^{-4} – 10^{-2}$
Solar Power Satellite (SPS) Wireless Power Transmission (WPT)	$10^{-1} - 10$
Solar radiation in the visible range	10^{2}

Photovoltaics versus Rectenna Technologies

When electromagnetic waves were experimentally observed, they were generated using antennas and radiating elements. Along the development of radio emission, antenna design became a separate area of expertise where the geometry of those elements configured the characteristics and capabilities of emission and reception of the EM waves. The shape and orientation of those antennas determine the polarization and direction of the emission, and reception. Electromagnetic spectrum was mastered and used in science and technology. Fortunately, the wavelengths associated with the radioelectric and microwave spectra allowed the manufacturing of radiating elements with the available fabrication tools. When increasing the frequency of the electromagnetic radiation, the geometries were shrunk accordingly and new fabrication strategies were used. Actually, an important leap in antenna design and fabrication appeared when using planar antennas written on flat substrates by microlithography techniques. Millimeter waves and Terahertz still benefit from those fabrication techniques. However, when the optical domain was placed as a feasible goal for antenna design, the use of electron beam lithography, focused ion beam and related nanometric precision manufacturing tools were necessary. Even more, those metals traditionally used as materials for antenna fabrication appeared to behave as non-perfect conductors, showing spectral dispersion and a non-negligible penetration depth.

At the same time that antennas were clearly devoted to the emission and detection of EM wave in the radioelectric, and microwave regimes, light and optical spectrum was covered with other reliable technologies for emission (incandescence lamps, spectral lamps, lasers, etc.) and detection (Golay cells, thermoconductors, photovoltaics, etc.) mainly based for detection in the quantified energy levels of semiconductors. Then, photodetectors improved their performance in responsivity, signal-to-noise ratio, cut-off frequency, size, and biasing requirements.

Then, it is easy to understand that antennas did not find a suitable place to develop as optical detectors. Semiconductor detectors were here to stay, fabrication of optical antennas is difficult and requires high-tech machinery, and metals are non-perfect conductors anymore in the optical regime.

However, some advances were made in using antenna-coupled detectors in the detection of light at higher and higher frequencies, and in its use as frequency mixer or coupled to bolometric devices. Besides, nanoscience has found optical antennas as promising elements to explore materials and media with high spatial resolution. Plasmonic optics has become an emerging field, where the collective oscillation of charges produces exotic phenomenologies that are used for sensing and probing sub-wavelength structures.

Several reports and papers [4, 5, 6, and 7] have been published in the past years presenting optical antennas and rectennas as harvesters of electromagnetic radiation in the infrared and visible spectrum. They are based on the principle of rectification of the currents generated in an antenna structure that resonates at the visible frequency. The idea, although appealing, has been somehow over-estimated when promising efficiencies above 80%. However, as we will see in this chapter, some important problems need to be addressed before fabricating an operative device. Unfortunately, the task of rectifying electric fields oscillating at 10^{14}-10^{15}Hz frequencies is formidable, and the efficiency figures obtained so far are well below the announced limit. The bottleneck of the technology remains in the rectification process. At the same time, some important advances have been made to tailor the impedance of optical antennas to properly couple the electromagnetic field and also to transfer the power to the load, i.e., to the rectifier. Then, optical rectennas can be considered as a promising technology with high potential. Based on the current results, more effort needs to be allocated to leap over the rectifying mechanism with novel technologies.

Although it is limited to the solar region of the electromagnetic spectrum, the most mature and standard technology (developed since the mid 70's) to harvest energy from EM radiation is photovoltaics (PV). According to the *National Renewable Energy Laboratory (NREL)*, conversion efficiency of PV technologies has been increasingly evolved during the last 40 years (figure 2). From the most simple variant of the 1st generation represented by the silicon based cells, to the 2nd and 3rd generations corresponding to thin-film and the most sophisticated multijunction cells respectively, a trade-off between efficiency and production cost is defining the market of each variant (table 5).

Table 5: Efficiency versus market of the 3 different PV technology generations

PV technology	Efficiency (%)	Market Share (%)
1st generation	20	90
2nd generation	5-12	10
3rd generation	40-50	--

The basic element in PV technology is the photovoltaic/solar panel/module which is composed of photovoltaic cells connected in parallel, when photogenerated current must be enhanced, or placed in series, when the output voltage is the parameter that needs to be maximized (see chapter 10: "Electronics for Power and Energy Management").

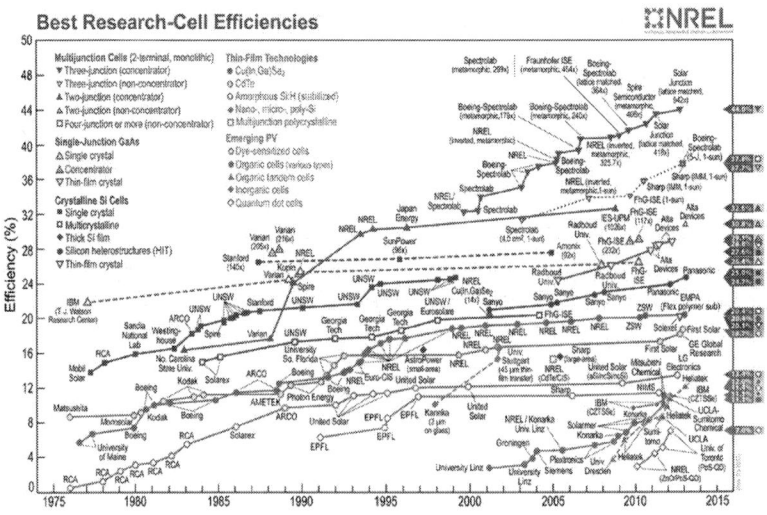

Figure 2: Efficiency evolution of the main photovoltaic technologies (from *National Renewable Energy Laboratory*).

The working principle of a photovoltaic cell is based on the photovoltaic effect, which was firstly described by Alexander-Edmond Becquerel in 1839. As it is reviewed in chapter 3 about Solar Energy Harvesting, the photovoltaic effect has the same quantum nature as the photoelectric effect, so both can only be described by considering that the energy of the electromagnetic radiation is quantized in quanta called photons, with an energy hv, as it has been explained before (equation 2). As it is shown in figure 3.a, photovoltaic effect takes place at the core of the cell, which is found at the junction of the two semiconductors that integrates a typical PV cell. When an individual photon interact with an individual electron at the valence band of the semiconductor, the energy of the photon (and the photon itself) can be absorbed by the electron to get promoted to the conduction band, leaving a hole in the valence band. This process called, photo-generation of an electron-hole pair, is only possible if the photon energy is at least equal to the energy of the band-gap (energy distance between the conduction and valence band). The population of photogenerated electrons and holes is then driven by the electrical field in the depletion zone of the PN junction and can eventually contribute to a photovoltage and the corresponding photocurrent, when an electric load is connected to the PV cell. In this case, both the photovoltage and the photocurrent are *dc* magnitudes and their product gives directly the electrical power converted by the PV cell.

Instead, the radiofrequency rectenna (RFR) technology is based on the combined operation of two basic elements: an electrical rectifier that follows an electromagnetic antenna (rectenna). The operation principle (figure 3.b) does not require quantum mechanics to be explained since, in this case, electrons in the metallic antenna are already in the conduction band, and do not need to be promoted in energy by absorbing photons from the electromagnetic radiation. In this case, the phenomenology is better explained by the interaction between the electrons in the antenna and the electric field of an electromagnetic incident wave. Similarly to PV technology, in rectenna technology matching conditions must be also satisfied. Now, the wavelength of the EM incident wave has to be a multiple of the antenna characteristic length in order to induce a resonant electrical current in the antenna. As opposite to a PV cell, an antenna will generate at its output both an *ac* voltage and an *ac* current. For this reason, a rectifier is needed as the first basic electrical component to transform *ac* values into *dc* values.

Optical rectenna (OR) technology can be considered as a particular case of rectenna technology where the frequency of the electromagnetic radiation involved is in the optical range. So, from this point of view, RFR technology covers the radiofrequency part of the electromagnetic spectrum and OR the optical part (figure 4). However, as it will be described in a next section of this chapter, OR technology cannot be considered just as an extrapolation of the RF rectenna concept to the optical range, since neither the antenna element, in this case a nanoantenna, nor the rectifier, typically a metal-insulator-metal (MIM) diode, have exactly the same properties of the RF counterparts. New physics such as plasmon resonances have to be taken into account in the optical antenna (OA), an antenna with characteristic lengths in the nanometer range (nanoantenna) to match the wavelengths of light radiation. Also special structures and materials are needed to achieve response times short enough to rectify signals in the THz range, which are induced in the nanoantenna element by the incident optical radiation.

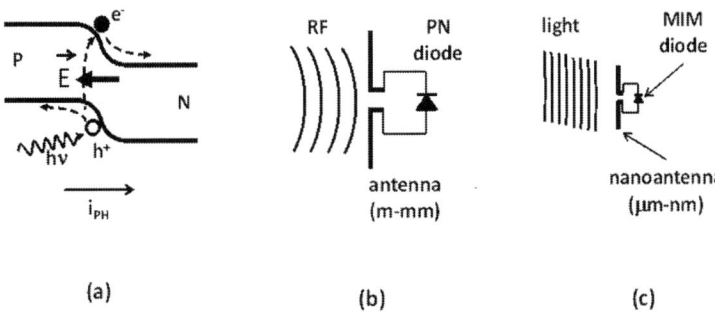

Figure 3: Scheme of the operation principle of three different technologies used to harvest energy from the electromagnetic spectrum (a) Photovoltaic technology (PV), (b) radiofrequency rectenna technology (RFR) and (c) optical rectenna technology (OR).

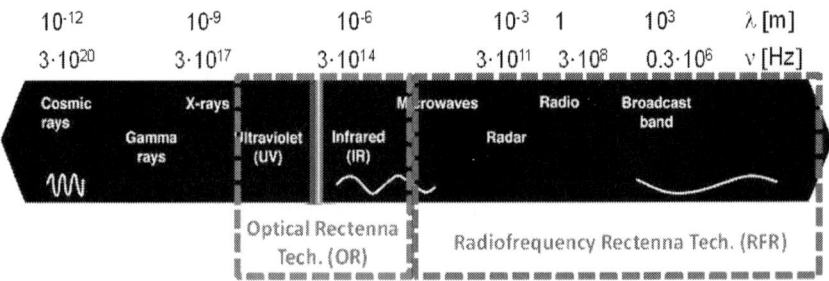

Figure 4: Spectral range coverage of the RF and optical rectenna technologies.

When used as light detectors, optical antennas involving rectifiers perform quite well in several specifications, especially in those related with their intrinsic electromagnetic nature. Table 6 shows these figures for a few technologies working in the visible and the infrared. We may already see in this table that the responsivity of optical antennas is lower than the rest of technologies. This figure is in accordance with the low efficiency of rectennas that has been observed in actual experiments involving MIM, or Metal-Insulator-Insulator-Metal (MIIM), junctions as transducers. Summarizing this table we may say that optical antennas are point detectors, very fast, work at room temperature, can be integrated with some other elements and devices (for example with focusing optics), and they present a broad tuneability, and a remarkable selectivity in direction and polarization.

Table 6: Four different photodetection mechanisms are compared with optical antennas technology

	Visible CCD/CMOS	MIM Junctions	Avalanche Photodetectors	Pyroelectric Detectors	Bolometric Detectors	Optical Antennas
Size	10^2 2	10^2 2	10^2 2	10^1-10^2 2	10^1-10^2 2	10^{-2}-10^0 2
Polarization selective, directivity, tuneability	No	No	No	No	No	Yes
Cooling	Better performance	No	Better performance	No	No	No
Responsivity	10^3-10^4V/W	0.7-0.9 A/W	0.7 A/W	10^3-10^4V/W	10^3-10^4V/W	0.1 V/W
Time response	100 ns	10 ps	9 ps	400 µs	400 µs	1 ps

Nowadays, space in urban areas, including work and home environments is strongly packed with EM radioelectric waves at various bands and spectral regions: besides the ubiquitous presence of radio and TV bands, cell phones and personal communications devices, a myriad of wi-fi stations, Bluetooth gadgets, and remote emitters and detectors produce a non-negligible amount of EM energy flowing around us. Then, from a harvesting point of view this energy could be recycled and properly used by electronic systems with ultra-low power requirements. This strategy may work in those environments with strong RF signals, where signal-to-noise ratio of other operative elements is not compromised. This idea of RF and microwave recycling has been developed some time ago in the form of antenna arrays and half or full wave rectifiers. Optical rectennas can be seen as an evolution and transposition of those designs and devices already working in the microwave region. In this band some designs have demonstrated more than 75% of efficiency when used for power transmission [8]. These figures are reduced when considering broad-band antennas designed to recycle microwave energy from ambient background.

Unfortunately, so far the efficiency number obtained at those frequencies have not been replicated at infrared or visible frequencies. The reasons are mostly derived from the inherent behavior of materials when frequency increases. Besides, the difficulties of designing THz electronics and oscillators, metals begin to behave as dispersive materials and the currents built on their surface penetrates within the structure.

In order to place the reader in a position to make an educated guess on the different technologies we present here a brief comparison among the photovoltaics, radiofrequency rectifiers, and optical rectennas (optical antennas coupled to rectifiers)

Photovoltaics: Direct conversion of light into electric power using the photovoltaic effect exhibited by semiconductor materials.

- *Efficiency:* The theoretical limit is around 41% for single junction solar cells, and reaches 87% for multiple junctions.
- *Pros:* Well established and mature technology. Fabrication issues have been solved due to the intrinsic relation with semiconductor technology.
- *Cons:* The performance is strongly dependent on temperature, especially for multiple junction cells.

Radiofrequency Rectifiers: Direct conversion of light into electricity using a rectifier working at radio or microwave frequencies.

- *Efficiency:* The limit is set around 85%. Practical devices have been demonstrated with efficiency larger than 75%.
- *Pros:* Well-known basic mechanism of rectification Fabrication can be made using standard photolithography on dielectric substrates.
- *Cons:* Polarization and spectral selectivity.

Optical Rectennas: Direct conversion of light into electricity using rectifiers working at optical frequencies.

- *Efficiency:* The theoretical limit is around 85%.
- *Pros:* Antenna theory and its scaling to optical frequencies is known and antenna-coupled detectors have been demonstrated in the infrared and the visible. Minimum size of about λ^2, allowing very high packaging. No dependence with temperature. Metals are used for fabrication with some advances in the use of conducting graphene.
- *Cons:* The efficiency of working devices is well below 85%. Barrier rectifiers are not able to follow optical frequencies and behave as square law rectifiers. Further advances are needed to have feasible rectifying mechanisms. Nano-fabrication technologies are necessary (nano-imprint could solve large scale fabrication numbers).

Historical Overview of Wireless Power Transmission

The precedents of the rectenna technology are found in the first attempts to transmit power through radio waves. The early history of RF power transmission dates from the experiments of Heinrich Hertz (1857-1894). Hertz was the first to rigorously prove the existence of electromagnetic waves. The experiments carried out for this purpose (figure 5) were based on transmitters and receivers of radio pulses that were combined with reflectors to create standing waves between the emitter and the receiver. In such experiments, *dc* power was converted to UHF radio waves by means of an LC oscillator connected to a device called *spark gap*. The emitted UHF EM wave from the dipole antenna was directed to the receiving antenna, an open loop ring with also a spark gap, by means of a parabolic reflector. When the emitted wave impinged on the receiving loop, a current was induced and a spark was produced. Hertz was in this way able to verify experimentally the existence and propagation in free space of EM radiated waves and to measure that such propagation is produced at the velocity of light.

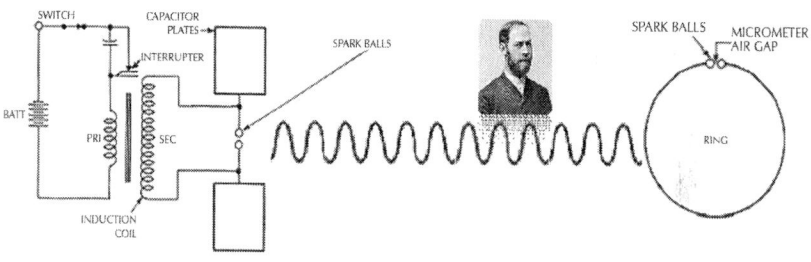

Figure 5: Scheme of the experiments carried out in 1887 by Heinrich Hertz to demonstrate the existence and propagation of electromagnetic waves in free space.

It was some years later, at the turn of the century, that Nikola Tesla (1856-1943) became interested in transmitting electrical power from one point to another wirelessly. Several famous attempts are described elsewhere. In the first one, which was carried out in 1899 at the Colorado Springs Laboratory, an approximately 60 m mast antenna with a 1 m diameter copper ball on top was built. An enormous coil that was fed with 300kW of electrical power got in resonance at a frequency of 150 kHz. When this coil was connected to the mast antenna, an RF potential of 100 MV with respect to the Earth was produced. The only record from this attempt were the discharges from the sphere to ground (figure 6left), but no data about the power radiated and the power collected at a certain point were reported.

Figure 6: Pictures of the Colorado Spring Laboratory experiments (left), Nikola Tesla (center) and Wardenclyffe plant (right).

Starting in 1901, a similar frustrated attempt to transmit power wirelessly was performed by Tesla in the Wardenclyffe plant at Long Island (figure 6 right), New York. In this occasion, a wooden tower 46 m tall was built to place a 30 m in diameter doughnut-like copper electrode. With this giant installation, Tesla wanted to transmit electrical power across the Atlantic, from the USA east coast to Great Britain. In 1914 the tower was demolished after Tesla lost funding to continue this project.

During the 1930's decade, less ambitious and more controlled experiments performed in the Westinghouse Laboratory led H.V. Noble to successfully transfer hundred watts of power between two 100 MHz dipoles placed 1.5m apart.

A retrospective analysis shows that the initial Teslas's failures and the lack of a clear demonstration of wireless power transfer during the first half of the past century is because power transfer starts to be efficient at the microwave range and above in frequency. At that time the technology to generate power in this wavelength range was not developed enough. It was not until the development of the klystron and the magnetron, combined with the end of the World War II that power transfer technology could start to be notably unfolded. A detailed description of this ramp-up period of the modern history of wireless power transfer is done by one of the most prominent protagonists, William C. Brown [9]. Brown is famous by the invention of the crossed-field amplifier, also known as *Amplitron*, but he can be also considered as the pioneer of the microwave power transmission and the first in developing a rectenna. Most of Brown's achievements were carried out at the Raytheon Company and at the Jet Propulsion Laboratory (NASA), and were mainly driven by two applications respectively: the Raytheon Airborne Microwave Platform (RAMP), a microwave-powered helicopter and the solar-power satellite (SPS), with microwave power transfer to the Earth. The requirements demanded by both applications gave rise to the development of the Amplitron and the first rectenna, as solutions to the generation of high Continuous-Wave (CW) powers of microwaves to be transmitted and to the direct conversion of the received microwaves into *dc* power, in order to drive the motors of the helicopter rotor blades. Thus, in 1964, the first flight of a helicopter prototype was demonstrated. It was propelled by the 270W *dc* provided by a 1.4 kg array-like rectenna integrated by 4480 1N82G semiconductor diodes in a 0.4 m² area, which corresponded to a power to mass ratio of 5 kg/kW. Eight years later, in 1968, the improvements introduced by the use of Schottky diodes produced an enhancement of this ratio of one order of magnitude. Finally, in 1983, the introduction of the thin film etched-circuit rectenna technology made possible *dc* to *dc* efficiencies of 85% and power to mass ratios of 1kW/kg.

Further advances in the wireless power transfer technology have been made during the last part of the past century and the beginning of the present one. As a consequence of this last evolution, a company called WiTricity Corp. and a technology called passive Radio Frequency IDentification (RFID) has become the two most meaningful examples of successful application of wireless power transfer.

WiTricity Corp. was born in 2007 to commercialize the applications of a technology developed at Massachussetts Institute of Technology (MIT) by Professor Marin Soljačić and co-workers [10]. The operation principle of this technology is based in the non-radiative power transfer between two self-resonant coils operating in the strong coupling regime.

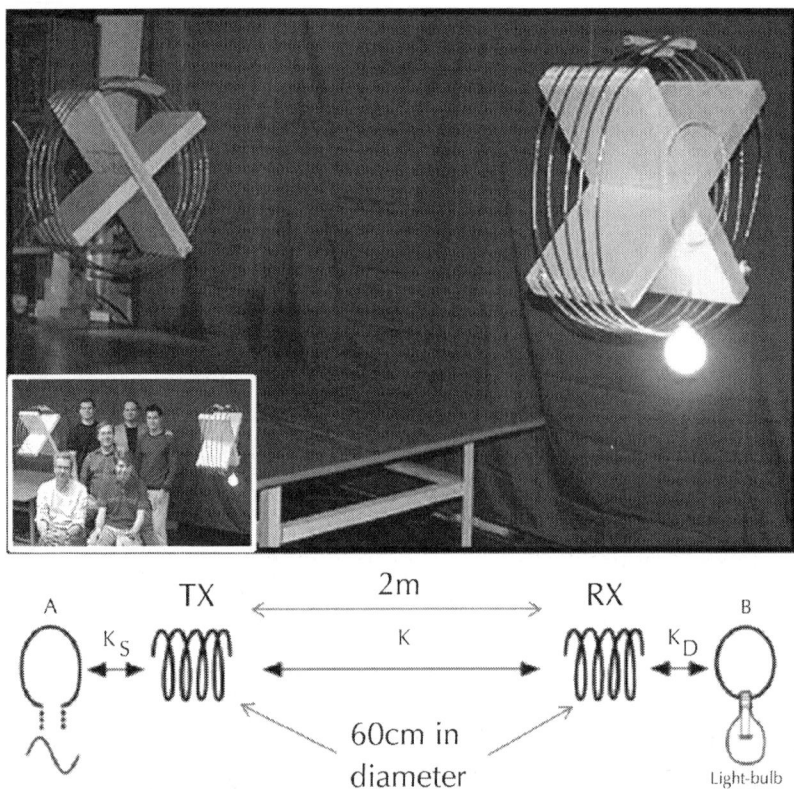

Figure 7: Picture (top) and scheme (bottom) of the WiTricity concept experimental setup. In the inset, the team of Prof. Soljačić at MIT is placed between both coils during operation, trying to demonstrate that the technology is harmless.

In figure 7, the setup used to demonstrate the WiTricity concept is shown. A single loop, A, connected to a sinusoidal signal generator is magnetically coupled to a secondary 5-turns emitter copper coil 60 cm in diameter. An identical receiver coil is coaxially placed at a 2m difference from the emitter, and also coupled capacitivelly to a secondary single loop connected to a 60W bulb load. The system, which is designed to resonate at 9.9 MHz, transfers the 60W of power needed to light the bulb with an efficiency of around 45%. At a shorter distance of 3 ft, the 60W are transferred with an efficiency of 90%.

Finally, RFID technology is also drawing on wireless power transfer technology [11]. In the active RFID variant, the RFID transponder, also called "tag", get the energy from a battery to supply its Application-Specific Integrated Circuit (ASIC). By contrast, in a passive RFID technology, the voltage generated in the tag antenna by the transmitted RF signal during the periods of unmodulated carrier is converted to a dc voltage. This voltage is used to power up the active ASIC chip circuitry which controls the input impedance of its front end. Communication between the base station (RFID reader) and the active tag is based on the modulation of the back-scattered signal produced by the toggle of the input front end impedance between two states (figure 8).

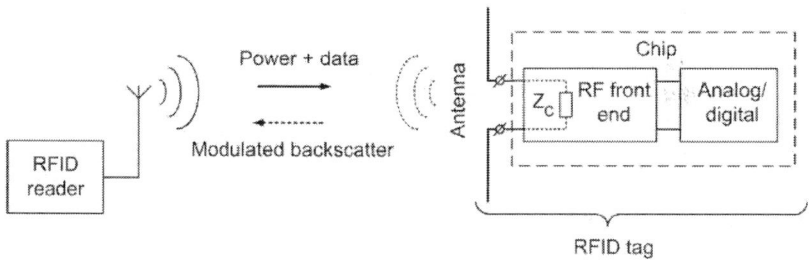

Figure 8: Scheme of an RFID system. When RFID is passive, the power transmitted by the reader during the unmodulated periods is converted in a dc power at the input of a passive circuitry of the tag to supply the rest of the active chip [11].

ELECTROMAGNETIC RADIATION ENERGY HARVESTING. THE RECTENNA APPROACH

As it has been pointed out above, a rectenna is the basic element of the RF and optical rectenna technologies. It basically consists (figure 9) of an antenna, in charge of efficiently collecting the energy emitted from a radiative source in the EM spectrum, and a diode, in charge of rectifying the ac voltage induced at the antenna terminals by the EM radiation. Eventually, a low pass filter follows the diode in order to obtain a dc voltage from the rectified signal. Usually, a dc-dc converter is also needed to adapt the voltage levels of the filter output with the level required by the application, represented infigure 9 by its equivalent load. As in most of the energy harvesters, control electronics will manage the flow of energy from the *dc-dc* converter to the application load or to a storage device, usually a battery, depending on whether the energy harvested by the rectenna can satisfy the application demand or, instead, it is better to store the harvested energy until the load demand could be satisfied.

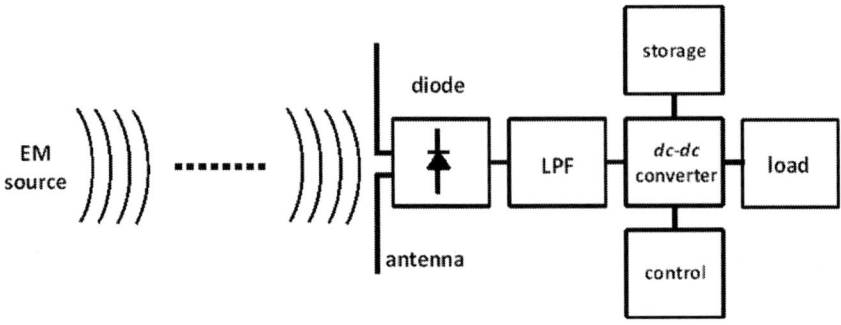

Figure 9: Block diagram of a rectenna.

The core of the rectenna, i.e. the antenna and the rectifier, can be replicated in an array configuration in order to improve the collection efficiency. Thus, a 2D array of identical rectenna elements can be connected in series or in parallel to increase the effective collection area and to increase the output voltage or the output current respectively [12]. Another 2D array configuration strategy is based on combining rectenna elements with different characteristics in order to match the different wavelengths of an EM sources set.

The Antenna as Transducer Element

An antenna is a device made to transmit and/or receive EM waves. By converting an electric current into an electromagnetic field on one end and converting this EM field into a voltage on the other, a pair of antennas gives the capability of making a wireless link between two points.

The most important parameters of a transmitter antenna are described below:

- Impedance: given the fact that the antenna must be connected to a transmitter, through a transmission line, and radiate the maximum amount of power with the lowest losses, the impedance of the transmitter, the transmission line and the antenna must be the same. The antenna itself introduces losses in the system, normally ohmic. In almost all the EM antennas the input impedance can be computed as the sum of the losses (R_Ω) and the radiation (R_r) resistance.

$$Z_{in} = R_r + R_\Omega$$

(8)

The total delivered power to the antenna should then be calculated as:

$$P_T = P_r + P_\Omega = I_{in} \cdot R_r + I_{in} \cdot R_\Omega$$

(9)

- Efficiency: given the losses and radiation resistances the efficiency of an antenna can be computed as

$$\eta_e = \frac{R_r}{R_r + R_\Omega}$$

(10)

- Directivity: the directivity of an antenna is defined as the relation between the radiated power density in one particular direction and the radiated power density which would radiate an isotropic antenna emitting the same power.

- Radiation pattern: the radiation pattern of an antenna describes the relative field strength of the radiated EM waves in all the directions from the antenna, at a fixed distance. For directional antennas the radiation pattern shows that there is a particular direction on which the antenna emits more efficiently. For omnidirectional antennas the radiation patter is almost equal for all directions.

- Gain: compared to an isotropic radiator, which will equally distribute the radiated power in all directions, real antennas with either a directional or omnidirectional radiation pattern will radiate less power in some directions and more in others. Therefore it can be considered that there is a gain between the different radiation directions. This gain can be defined as the ratio between the transmitted signal strength value at the more efficient direction and the value using a reference antenna. If the reference antenna is an isotopic source the used units will be dBi.

- Polarization: the polarization of an antenna is defined as the orientation plane in which the radiated or absorbed electric field vibrates with respect to a reference plane, for example the Earth's surface. It is determined by the physical construction of the antenna and its orientation, especially by its radiating element. The most common polarizations are linear and circular, which are particular cases of elliptical polarization. In the first one, the electric field vector stays in the same plane whereas for the second one it appears to be rotating with a circular motion around the direction of propagation.

- Bandwidth: most of the EM antennas operate efficiently over a relatively narrow frequency span due to their geometry. As a consequence, they must be tuned in order to have the same frequency band operation as the electronic system at which they are connected.

The most important parameter of a receiver antenna is

- The effective area: an antenna extracts power from the wavefront of an EM wave, thus it represents a certain capture area or effective area. This area is defined as the relation between the powers delivered to the receiver circuitry and the power density of the incident wave.

Additionally, antennas comply with the law of reciprocity. This law states that given two identical antennas placed at some distance, each of them can be operated either as a transmitting antenna or as a receiving antenna. Suppose that the one working as a receiver is kept intact, while the performance is modified so that, for a fixed amount of radiated power, the signal received by the other antenna changes by a factor. If the same modified antenna is used for receiving the transmitted signal by the unmodified one, its performance will also be changed by the same factor. This theorem can be formally derived from Maxwell's equations and its validity can be easily verified.

As a consequence of this law, all the previous described antenna characteristics (efficiency, radiation pattern, gain, polarization, bandwidth and effective area) are the same whether the antenna takes part in a transmitter or a receiver scheme. Besides, when designing optical antennas, this reciprocity law is used to simplify the calculation, for example, when calculating the radiation/receiving patterns and some other important parameters of the antenna.

The simplest designs of radiofrequency antennas are developed as wires or loops of conductors properly arranged and connected to an electronic element to produce and detect electromagnetic radiation with a wavelength scaled to the size of the antenna. An important step forward was made when planar structures used the resonant properties of metal patches and planar strips. These could be fabricated using photolithography in a very similar manner as it was done with printed circuits. When the frequency increases the resolution of manufacturing techniques also increases to produce thinner and finer structures. Then, when moving to terahertzs and infrared frequencies only nanofabrication techniques are suitable to realize those antennas. The reasons are twofold. On the one hand is the shrinking of the wavelength towards the nanometric scale, and on the other hand the quality of the finishing elements in terms of roughness and surface smoothness, which may interact with the currents and scatter the charge carriers around non useful directions. These nanophotonic devices are still considered as antennas because they can produce or detect electromagnetic radiation using wires or resonant patches. In a simple manner, optical antennas are defined as resonant structures able to produce an electric signal related with the incident optical radiation.

At the same time, taking into account the re-emission of electromagnetic radiation by resonant structures, it is possible to define a new kind of element that changes the properties of the light that interacts with it. We will name these elements as belonging to the "resonant optics" area. Conventional, reflective or refractive, and diffractive optics relies on the geometrical and wave models of light. Then, resonant optics uses the electromagnetic interaction of light waves with geometrical structures, typically fabricated with conductors, which work building currents up. They have been used as frequency selective surfaces, polarization elements, or phase shaping devices. Some of the designs are scaled versions of their microwave counterparts, where they were first demonstrated. In the recent past we have seen a growing number of scientific contributions where these elements are analyzed and exploited for a variety of applications. At the same time, they have been denoted with different names, mainly depending on the origin of the research teams that develop them: metamaterial surfaces, flat optics, 2.5D photonic crystals, etc.

Both optical antennas and resonant structures are determined by geometry and material parameters. Geometry mostly drives the polarization and spectral selectivity. Intrinsically, the size of the antenna is related with the wavelength. Therefore, when considering infrared and optical radiation, optical antennas become, by nature, nanophotonic devices. For example, a dipole antenna has a length of a few microns for far infrared radiation, and a few hundreds of nanometers for visible light. At the same time, the width of the dipole is limited by fabrication constrains and can be as narrow as a few tens of nanometers. Consequently, the area of detection of the incoming radiation, which extends a little farther from the antenna itself, is about l^2, depending on the geometry of the antenna. The far-field pattern of optical antennas resembles that of their radiofrequency counterparts.

Also, as it happens with planar antennas written on dielectric substrates, the responsivity of optical antennas is larger when light is incident from the substrate side than when it is incident from the air, mostly because of the larger electric permittivity of the substrate. The ratio between the power radiated, or received, by an antenna located between two media of electric permitivities ε_1 and ε_2 is [13, 14]

$$\Gamma = \frac{P_1}{P_2} = \left(\frac{\varepsilon_1}{\varepsilon_2}\right)^{3/2}$$

(11)

Besides, the effective wavelength at which the resonance takes place moves because of this situation. This effective wavelength is given classically as

$$\lambda_{eff} = \frac{\lambda_0}{\sqrt{\frac{\varepsilon_1 + \varepsilon_2}{2}}}$$

$$(12)$$

And when considering the plasmon resonances as

$$\lambda_{eff} = n_1 + n_2 \frac{\lambda_0}{\lambda_p}$$

$$(13)$$

Where n_1 and n_2 depend on the geometry and material parameters and λp is the wavelength of the plasmon resonance [15]. This equation already shows the influence of the material parameters in the performance of optical antennas. At the infrared and visible frequencies metals are no longer perfect conductors and behave as dispersive materials [16]. This means that the radiation losses increase and the surface currents penetrate deeper within the materials. However, metals also present interesting phenomena at optical frequencies. The optical radiation can excite the collective resonance of the charge carriers. They now, oscillate as a unique particle that is named as plasmon. The plasmonic resonances have been devoted increased attention along the past years, producing a myriad of papers and novel applications. In the area of optical antennas, plasmons play an important role because of their occurrence at optical frequencies.

Although in this chapter we focus our attention on optical antennas, we have to mention the important role that resonant structures may play in the improvement of photovoltaic solar cells. It is known that when populating a surface with metal nano-structures, the interaction of light with the structure changes. If the surface is that one of a solar cell, photons can be scattered by the resonant structures and the optical path within the material is enlarged. Also the enhancement of the electric field near the nano-structures may increase absorption, or directly, those photoexcited electrons can be injected into the cell, contributing to the total current delivered by the cell [17]. Some advances have been made reporting 50% increase of the transmittance of the surface using plasmonic nanoparticles [18]. This path is quite promising for improving the performance of traditional photovoltaic cells.

Rectifying Devices and Technologies

Rectification is commonly performed by p-n junction diodes when RF radiation is in the kHz-MHz low frequency range. However, when operation frequencies are in the GHz-THz range, semiconductors and devices with shorter transit times and lower intrinsic capacitances like GaAs Schottky diodes are needed. Typical maximum operation frequency of Schottky diodes is 5 THz and although theoretical efficiencies approach 90%, only values of just around 50% have been demonstrated experimentally.

In most common rectifying situations, which correspond to low frequency and high power conditions (LFHP), diodes produce a half-wave rectification with an efficiency given by

$$\eta_{LFHP} = \frac{1}{1 + \frac{v_D}{2v_{dc}}}$$

(14)

Where v_D and v_{dc} are respectively the voltage drop across the diode and the output rectified *dc* voltage.

However, in the opposite conditions of high frequency and low power (HFLP), which corresponds to those of RF signals to be harvested by a rectenna, rectification is much more complex Since: first the incident power is not only low but fluctuant in value and, second, because matching of the antenna and the diode is an issue. Consequently, the power at the output of the rectenna, P_{dc}, will be obtained from the incident RF power, P_{RF}, by

$$P_{dc} = P_{RF} \cdot \eta_{HFLP}(P_{RF}, \rho)$$

(15)

Where the rectification efficiency, η_{HFLP}, depends explicitly on P_{RF} and on the antenna to diode matching ρ this explicit dependence cannot be described analytically with a closed expression like in the LFHP case and the problem of P_{dc} prediction has to be solved by simulation. Time-domain analysis has been successfully applied to address this problem for single frequency or narrow-band rectenna applications [19]. However, a frequency-domain approach based on the harmonic balance (HB) method is more appropriated for wide-band applications [12], where some characteristics of the diode as the nonlinearity of its capacitance, the reflected harmonic energy at input/output or the self-biasing effects start to be relevant.

In the case of optical antennas, they work by combining the effect of two physical mechanisms. One of them is the coupling of the optical radiation to the device. This task is in charge of the metal structure. The other is the transduction mechanism used to provide the output signal. So far, two main types of transducers have been demonstrated. The bolometric response of the material is used to produce a change in the voltage measured from the device. However, this mechanism is dissipative and it does not provide a positive balance of power, and therefore it is not useful for harvesting applications, at least directly. As an interesting outcome of the Joule dissipation we find that when incorporating resonant elements to thermoelectric pairs, the combination of localized heating and the Seebeck effect can be of use to produce electric power [20, 21]. The other mechanism is the rectification of currents using a diode. This is the case implemented in rectennas. Metal-Insulator-Metal (MIM), and Metal-Insulator-Insulator-Metal (MIIM) have been demonstrated as effective diodes from the microwave to the visible wave ranges [22, 23, 24, and 25]. Typical MIM materials are $Cr/CrO_x/Au$, $Nb/NbO_x/Nb$ or $Al/AlOx/Pt$, and state of the art MIM diodes can operate at frequencies up to 150 THz. These junctions work as square-law rectifiers. The rectified current is given as

$$I_{DC} = \gamma \, \frac{|V_{diode}|^2}{4R_{diode}}$$

(16)

Where γ represents the non-linearity of the current-voltage curve of the device Non-linearity, γ, which is defined from the i-v diode characteristic as $\gamma \equiv (d^2i/dv^2)/ (di/dv)$, should be at least 3 or larger to start getting reasonable values of conversion efficiency Figure 10 shows the line transmission schematics of the antenna-diode element. When optimizing this structure for maximum efficiency the impedance of the antenna has to compensate the impedance of the load (the diode). This means that the antenna has to present an imaginary part of the impedance that is not typically included when maximizing the performance.

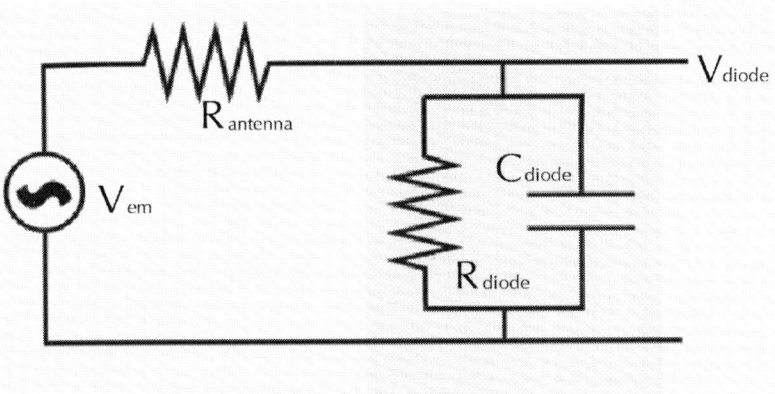

Figure 10: Transmission line schematics of a diode coupled to an antenna. The antenna is considered as having an impedance of $R_{antenna}$.

On the other hand, the cut-off frequency of the element is given as

$$f_{cutoff} = \frac{1}{2\pi RC}$$

(17)

Where the RC constant should be smaller than 10^{-14} s^{-1} to reach the infrared and optical regions The junction itself works as a capacitor. To reduce its capacitance we cannot increase the thickness of the junction because in that case tunnelling would be not possible. Because of that, it has a thickness of about 2-3 nm, depending on the insulator. Therefore, in order to have a low capacitance the junction area should be small. On the other hand, to efficiently transfer the power to the load, the impedance mismatch should be corrected. These issues are being addressed in several ways and some promising results have been already published [26, 27]. However, when combining MIM diodes with antennas, the conversion efficiency given as the ratio between the power obtained after rectification and the incident power is quite low for these square law rectifiers [28]. The values for this efficiency is lower than 10^{-6}, showing that some better rectifying technologies need to be developed before practical devices based on direct rectification of light, become competitive against other solar energy harvesters. In Figure 11 we show the response of an optical antenna placed in front of a black body at 1000°K, demonstrating the existence of an output signal for this extreme condition.

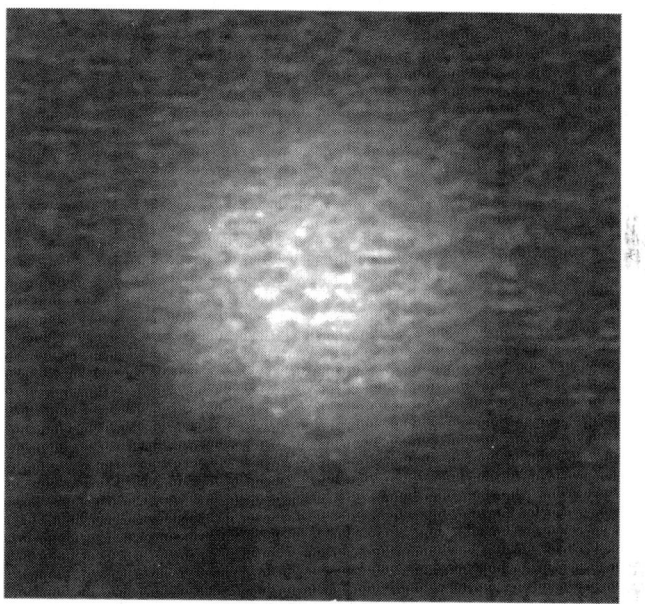

Figure 11: Response of an optical antenna located on the image of a black-body radiator at 1000°K. This image was obtained by a relay optical system fabricated in ZnSe and having F/ # equal to 1 [29].

A modification of the traditional MIM diode, the travelling-wave metal-insulator-metal diode (TW-MIM), allows obtaining a quantum efficiency of 3.6% in the IR region [30]. The TW-MIM is based on the rectification of the surface plasmon excited by the antenna on a plasmonic waveguide.

Some interesting advances in rectification at terahertz and quasi-optical frequencies have been proposed using a novel approach. The geometric rectifier (see Figure 12) has been demonstrated at GHz frequencies [31]. Here the rectification is given by deflecting the trajectories followed by the charge carriers through an asymmetric channel. The simplest case is an arrow shaped element that selectively directs charge carriers in a given direction, preventing the movement of those carriers towards the opposite direction [32]. These geometric rectifiers need materials and conditions where the carriers exhibit a free mean path longer, or much longer, than the size of the rectifying structure, i.e., a few hundreds of nanometers. Besides the electric properties of metals, exhibiting free mean paths in the range of a few tens of nanometers (for example for Au this parameter is around 20 nm), the conduction properties of graphene can be tailored to produce feasible devices having this effect [33].

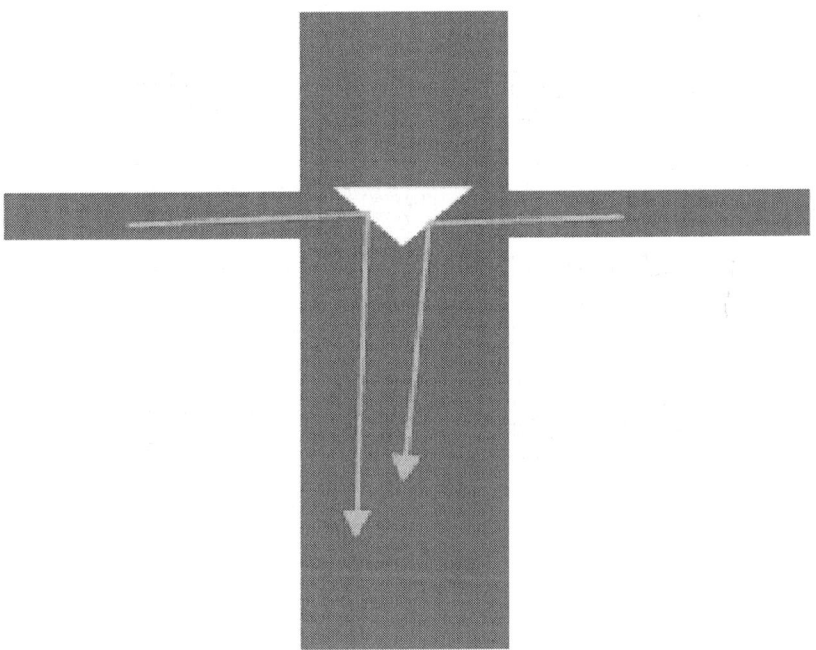

Figure 12: Basic scheme of a geometric rectifier. The horizontal arms of a dipole antenna intersect at the feed point. This feed point is shaped as an asymmetric defect that deflects charge carriers towards the bottom of the geometric rectifier [34].

Also a device based on grapheme, the field effect transistor (G-FET) in common source configuration, can potentially be a promising candidate as rectifier component for optical rectennas. The extremely high mobility of graphene combined with the ambipolar transport properties would allow to implement full-wave rectification at THz in a single device [35].

Finally, two examples of RF rectenna (RFR) and optical rectenna (OR) from the literature are shown infigure 13 in order to compare dimensions and performances.

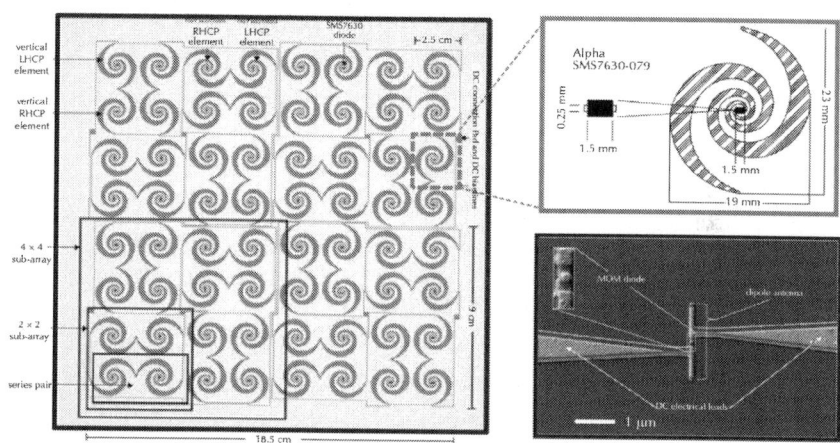

Figure 13: RF rectenna array (left) integrated by 64 single spiral rectennas (right top) [12] Optical rectenna for the IR range consisting on a dipole coupled to a MOM diode (right bottom) [36].

The RF rectenna of figure 13 corresponds to a rectenna array configuration designed to operate in the 2-18 GHz region and for input power densities from 10 nW/cm² to 0.1 mW/cm². The spiral rectenna elements are distributed along the 324 cm² array area with different orientations in order to harvest energy from different polarized sources. So, considering an effective area of A_{eff}=25 cm², then input RF power will vary between P_{RF}=250 nW and P_{RF}=2.5 mW. If rectification efficiencies for such input powers results to be η (250nW)=1% and η(2.5mW)=20%, then the output dc power harvested by this array will theoretically vary between 2nW and 450 μW. Other examples of rectenna designs achieved efficiencies from 40-50% operating at 10^{-2} mW/cm² [37] to 80% at 10 mW/cm² [38].

By contrast, in the optical rectenna (figure 13 right bottom), designed to operate in the IR band, the 19x23 mm² area of the RF rectenna element is reduced to less than 1 μm^2. In this case, an $Al/AlO_x/Pt$ MOM diode is chosen to implement the rectifier, which is coupled to a dipole 1 μm long nanoantenna. However, as it has pointed out above, the maximum conversion efficiency of 20% shown by the RF rectenna is reduced below 10^{-6} in optical rectennas like this one.

CONCLUSIONS

In this chapter, the most important concepts needed to understand how energy from the electromagnetic spectrum can be harvested by means of the rectenna technology have been introduced. Main differences with the well stablished photovoltaic approach have been analyzed and a comparative list of pros and cons has been provided. An historical overview of the first works on wireless power transmission has been useful to understand the origin of the rectenna concept. The most relevant technical characteristics of both components of a rectenna, the antenna and the rectifier device, have been also decribed, and the specific features of each element have been explained for the readiofrequency and the optical range.

ACKNOWLEDGEMENTS

This work has been partially supported by project ENE2009-14340 from the Ministerio de Ciencia e Innovación of Spain and FP7-EU-Project FET-Proactive CA no.270005 (ZEROPOWER).

REFERENCES

1.   Shevgaonkar, R. K. *Electromagnetic waves*. Tata McGraw-Hill Education, 2005.

2.   Recommendation, Council. "519/EC of 12 July 1999 on the limitation of exposure of the general public to electromagnetic fields (0 Hz to 300 GHz)."Official Journal L 197 (1999): 1999.

3.   IEEE Std C95. 1-2005. "IEEE Standard for Safety Levels with Respect to Human Exposure to Radio Frequency Electromagnetic

Fields, 3 KHz to 300 GHz." New York: The Institute of Electrical and Electronic Engineers, 2005.

4. B. Berland, "Photovoltaic Technologies Beyond the Horizon: Optical Rectenna Solar Cell", Report for the National Renewable Energy Laboratory, NREL/SR-520-33263 (2002)

5. R. Corkish, M. A. Green, T. Puzzer, "Solar energy collection by antennas", Solar Energy, 73, 395-401 (2002).

6. D. K. Kotter, S. D. Novack, W. D. Slafer, P. J. Pinhero, "Theory and manufacturing Processes of Solar Nanoantenna Electromagnetic Collectors", Journal of Solar Energy Engineering, 132, 011014 (2010).

7. G. A.E. Vandenbosch, Z. Ma, "Upper bound for the solar enegy harvesting efficiency of nano-antennas", Nano Energy, 1, 494-502 (2012).

8. B. Strassner and K. Chang, "A circularly polarized rectifying antenna array for wireless microwave power transmission with over 78% efficiency," in*IEEE MTT-S Int. Microwave Symp. Dig.*, (2002), 1535–1538, (2002).

9. W.C. Brown, the *History of Power Transmission by Radio Waves*. IEEE Transactions on Microwave Theory and Techniques, 32 (9), 1230-1242 (1984).

10. A. Kurs, A. Karalis, R. Moffatt, J. D. Joannopoulos, P. Fisher, M. Soljac̆ic, "Wireless Power Transfer via Strongly Coupled Magnetic Resonances", Science 317, 83-86 (2007).

11. K. V. Seshagiri Rao, P. V. Nikitin, and S. F. Lam, "Antenna Design for UHF RFID Tags: A Review and a Practical Application", IEEE TRANSACTIONS ON ANTENNAS AND PROPAGATION, 53, 12, 3870-3876 (2005).

12. J. A. Hagerty, F. B. Helmbrecht, W. H. McCalpin, R. Zane, Z. B. Popovic, "Recycling Ambient Mricowave Energy With Broad-Band Rectenna Arrays", IEEE Transactions on Microwave Theory and Techniques, 52, 1014-1024 (2004).

13. C.R. Brewitt-Taylor, D.J. Gunton, H.D. Rees, Planar antennas on a dielectric surface, Electron. Lett. 17. 729–730, (1981).

14. J. Alda, C. Fumeaux, M. Gritz, D. Spencer, G. Boreman, "Responsivity of infrared antenna-coupled microbolometers for

air-side and substrate-side illumination", Infrared Physics and Technology, 41, 1-9 (2000).

15. L. Novotny, "Effective wavelength scaling for optical antennas", Physical Review Letters, 28, 262602 (2007).

16. J. González, J. Alda, J. Simon, J. Ginn, G. Boreman, "The effect of the metal dispersion on the resonance of antennas at infrared frequencies", Infrared Physics and Technology, 52, 48-51 (2009).

17. C. Hägglund, M. Zäch, G. Petersson, B. Kasemo, "Electromagnetic coupling of light into a silicon solar cell by nanodisk plasmons", Applied Physics Letters, 92, 053110 (2008).

18. P. Spinelli, M. Hebbink, R. de Waele, L. Black, F. Lenzmann, A. Polman, "Optical Impedaance Matching Using Couple Plasmonic Nanoparticle Arrays", Nanoletters, 11, 1760-1765 (2011).

19. T. Yoo and K. Chang, "Theoretical and experimental development of 10 and 35 GHz rectennas," IEEE Trans. Microwave Theory Tech., 40, 1259–1266, (1992).

20. C. Fu, "Antenna-coupled nanothermopile", M.S. Thesis, University of Central Florida (1998).

21. G. P. Szakmany, P. Krenz, A. Orlov, G. Bernstein, W. Porod, "Antenna-Coupled Nanowire Thermocouples for Infrared Detectioin"; Proc of the 12th. IEEE International Conference on Nanotechnology (2012).

22. A. Sanchez, C. F. Davis, Jr., K. C. Liu and A. Javan, "The MOM tunneling diode: theoretical estimate of its performance at microwave and infrared frequencies," Journal of Applied Physics. 49, 5270-5277 (1978).

23. C. Fumeaux, W. Herrmann, F. Kneubühl and H. Rothuizen, "Nanometer thin-film Ni-NiO-Ni diodes for detection and mixing of 30 THz radiation," Infrared Phys. Technol. 39, 123-183 (1998).

24. B. J. Eliasson, "Metal-Insulator-Metal Diodes for Solar Energy Conversion," Ph.D. Thesis, University of Colorado (2001).

25. C. Fumeaux, J. Alda, G. Boreman, "Lithographic antennas at visible frequencies", Optics Letters, 24, 1629-1631 (1999).

26. Jer-Shing Huang, Thorsten Feichtner, Paolo Biagioni, and Bert Hecht, Impedance Matching and Emission Properties of Nanoantennas in an Optical Nanocircuit" NanoLetters, 9, 1897-1902 (2009).

27.   Yauhen Sachkou, Andrei Andryieuski, Andrei V. Lavrinenko, "Impedance Conjugate Matching of Plasmonic Nanoantenna in Optical Nanocircuits", Proceedings of the 53rd International Symposium ELMAR-2011, 389-391 (2011).

28.   E. Briones, J. Alda, F. J. Gonzalez, "Conversion efficiency of broad-band rectennas for solar energy harvesting applications", Optics Express, 21(S3), A412-A418, (2013).

29.   J. Alda, "Response of an optical antenna to blackbody radiation", Personal Communication to G. Boreman (April, 1999).

30.   S. Grover, O. Dmitriyeva, M. J. Estes, and G. Moddel, Traveling-Wave *Metal/Insulator/Metal Diodes for Improved Infrared Bandwidth and Efficiency of Antenna-Coupled Rectifiers*, IEEE TRANSACTIONS ON NANOTECHNOLOGY, 9 (6), 716-722, (2010).

31.   A. M. Song, ``Electron ratchet effect in semiconductor devices and artificial materials with broken centrosymmetry", Applied Physics. A, 75, 229-235 (2002).

32.   Garret Moddel, "Geometric diode, applications and method", US Patent Office, Pat. No.: 20110017284 (2011).

33.   Z. Zhu, S. Joshi, S. Grover, and G Moddel, "Graphene Geometric Diodes for Terahertz Rectennas," J. Phys. D: Appl. Phys. in press (2013).

34.   J. Alda, "Geometrical Rectification", Personal Communication to G. Boreman (August, 2006).

35.   H. Wang, D. Nezich, J. Kong, and T. Palacios, *Graphene Frequency Multipliers*, IEEE ELECTRON DEVICE LETTERS, 30 (5), 547-549 (2009).

36.   J. A. Bean, A. Weeks, and G. D. Boreman. "Performance Optimization of Antenna-Coupled Al/AlOx/Pt Tunnel Diode Infrared Detectors", IEEE JOURNAL OF QUANTUM ELECTRONICS, 47, 126-135 (2011).

37.   W. C. Brown, *An experimental low power density rectenna* in IEEE MTT-S Int. Microwave Symp. Dig., 197–200 (1991).

38.   J. O. McSpadden, F. E. Little, M. B. Duke, and A. Ignatiev, *An in-space wireless energy transmission experiment*, in Proc. IECEC Energy Conversion Engineering Conf., 1, 468–473 (1996).

Chapter

6

Features of Usage of Electromagnetic Field of Extremely Low Frequency for the Storage of Agricultural Products

Kasyanov Gennady Ivanovich[1], Syazin Ivan Evgenyevich[1], Grachev Alexandr Vasilyevich[1], Davidenko Taisiya Nikolaevna[2], and Vazhenin Evgeniy Igorevich[1]

[1]Meat and Fish Products Department, FSBEI HPE, Kuban State Technological University, Krasnodar, Russia; [2]Phylology Department, FSBEI HPE, Kuban State Technological University, Krasnodar, Russia

ABSTRACT

The way of increase of storage period of agricultural raw materials under the influence of low frequency electromagnetic field (EMF LF) has been considered in the article. Existing developments of the EMF LF usage on the base of patents have been examined. The results of EMF LF processing of wood, wine, seeds, vegetative, fish and meat products have been presented.

ANALYSES OF EXISTING DEVELOPMENTS

The magnetic field as an active factor in agriculture has more than 50-year history. Important applied investigations were made in Russia, the results of which were the creation of technological lines of magnetic systems for pre-processing vegetative plantings seeds. Conditions under which the application of the magnetic field shows stable results stimulation—increase of the yield by 12% - 30%, depending on the culture, the initial quality of the seeds and growing conditions of plants—were found out.

The main application results with the help of gradient magnetic field (GMF) application have been obtained. Low energy consumption of GMF-processing of seeds in combination with a low intension of magnetic field (1 - 12 mT), absence of noise magnetic field outside the zone of seeds processing (a level of tension equals 0.01 mT at a distance of 0.5 m from the processing zone) makes these plants absolutely safe for service personnel. Nonhigh price of GMF-plant complements the advantages of this equipment.

Almost all the scientific works devoted to the impact of electromagnetic field of EMF LF or of GMF, were intended to stimulate the vitality of plants.

The exposure of extremely low frequency electromagnetic field having certain parameters has the property to suppress the growth of fungal microorganisms and bacteria almost completely [1].

The biological effects of modified water with isotopic composition, magnetically processed water and electrochemically activated water were examined [2]. It was shown that the light water can be used to create functional products. The electro-activated water with different pH and the redox potential has been obtained. The relaxation period and the return to the stable condition of water samples have been investigated. It was determined that the addition of alkaline fraction of activated water for the salting of chopped meat let to move the pH to higher values of the isoelectric point of the muscle proteins, which in sausage production can be made by the entering of various phosphates.

The way of agricultural products storage that provides the laying of beet piles has been developed [3]. While the storing before the processing, the roots have been processed by a constant magnetic field (which magnetic induction was 0.1 mT) during 50 min, and the amplitudemodulated electromagnetic field with a carrier frequency of 1 kHz and the frequency modulating oscillations, which are in the low frequency range of 3 - 30 Hz. The method reduces the loss of sucrose in the roots during the storage and increases its output by processing them in a sugar factory.

The way of beet storage in beet-sugar factories has been developed [4]. The method involves laying beets in piles next to it and location close to laser and radiator of electromagnetic field so that the maximum number of lines of the magnetic induction of electromagnetic field permeating beets. During the storage before reprocessing an exposure by extremely low electromagnetic field in the range of 3 - 30 Hz for 40 min with the following processing by the laser radiation in the range of 610 - 680 nm began in roots. The method let to reduce the loss of sugar in the roots during the storage and enhance its release in the beet reprocessing plant.

The method of agricultural seeds processing before seeding which increases the germination of seeds and the performance of the production line has been developed [5]. The seeds before planting are processed by the static magnetic field at its strength of 200 to 900 A/m and at the same time by the electromagnetic field, phase-modulated waves of extremely low range during 40 - 60 min at a field strength of 120 - 1400 A/m.

The way of drying has been developed [6]. The invention relates to the wood and the food industries, in particular, to drying methods of wood and of vegetative and animal raw materials origin. The method consists of the fact that the drying process of wood, fish, meat, cherries, carrots, etc. produced by EMF LF ranges of various modifications, with the carrier frequency for all types of modulated electromagnetic fields can be 30 - 1013 Hz, and the intensity of each of electromagnetic field is 10 - 1400 A/m, the length of exposure is 10 - 900 min. This method reduces the power consumption in the comparison with known methods down to 10^3 times.

The importance of discovered abilities is difficult to overestimate. There are tremendous opportunities to regulate the development of fungi in the case of safe preservation of agricultural products, for example, in wining to stop the fermentation process or usage in the production of animal feed, etc.

The action of the electromagnetic field (EMF LF) on biological systems was studied almost since the first generators of EMF LF. The exposure of electromagnetic fields on biological systems in the significant tension (leading to thermal effects in biological systems) has been studied sufficiently on this moment, but the weak impact of low-intensity fields (non-thermal nature) has not been sufficiently studied [1].

In an investigation of weak interactions, the traditional approach of describing the object of quantitative characteristics encounters fundamental and insurmountable difficulties. Quantitative characteristics of the exposure (for example, the coherence or an order) have an essential meaning in this sphere [7]. In addition, under the influence of the first type, biological objects usually activate protective mechanisms which help to compensate this effect, but weak influence doesn't have such effect [8].

The theme of specific action of EMF LF (3 - 30 Hz) was described in many scientific works, especially increased in the last 10 years, but the primary mechanism of this action is unclear to the end till this day [9].

Having based on the results obtained in the sphere of biophysical investigations, it is possible to develop new low-power-consumption methods of processing of fish raw materials with the aim of the reduction of microbial contamination and the shelf life extension. Electromagnetic waves of different ranges propagate in the space differently. Difference in the conditions of electromagnetic waves propagation is caused primarily by the fact that at different frequencies, the ratio changes between the displacement current and the conduction current environment.

THE PURPOSE OF INVESTIGATIONS

The aim of the investigation was to increase the nutritional value and the commercial quality of agricultural raw materials by processing by impulse EMF LF of the seminal material and laid deposit products.

RESEARCH METHODS AND LABORATORY EQUIPMENT

In this investigation the authors used the device "IMP-04" to measure electromagnetic environment parameters. This device usually is applied to measure the magnetic flux density of electromagnetic fields produced by the EMF LF generator. The frequency ranges of the measured signals of the device were the next: from 5 Hz to 2 kHz— "Band I", and 2 - 400 kHz—"Band II".

An important part of the experiment was to use the generator signals LF G3-122, destined for processing of agricultural raw materials in the frequency range 0.001 - 2 × 106 Hz (in increments of 0.001 Hz). The range of adjustment of the output voltage (R = 50 ohms) was 2 × 10 - 6 – 2.5 V.

The main parameters of EMF LF are the wave length (λ) and the frequency (f) which is related to the back wavelength dependence (for the conditions of the wave in the air):

$$f = c/\lambda,$$

(1)

Where c is speed of light.

The scheme of processing of liquid environment by EMF LF is presented in Figure 1.

One of the most important ways to reduce losses of vegetable production is the usage of rational storage methods. Existing methods of vegetables storage are well studied. Some of them have wide application both in Russia and abroad (for example, the bulk potato storage with active ventilation), but the usage of the other (Modified Gas Atmosphere, Regulated Gas Atmosphere) is limited because of the high costs of their implementation. Along with the basic methods, additional ones of increase of the vegetable products preservation (bleach, growth-inhibiting, with surface protective package) are used. Many of the additional storage methods are under study and are debatable. Their efficiency is not confirmed yet in the manufacture. The explore of the possibility of the pulsed low-frequency electric field in the purpose to improve the nutritional value, the quality and the persistence of commercial vegetable production is a very important task which determined the choice of research.

The analysis of the data has shown that the resonance impact effect of EMF LF on the studied parameters is observed at frequencies of 19.52 and of 40.03 Hz.

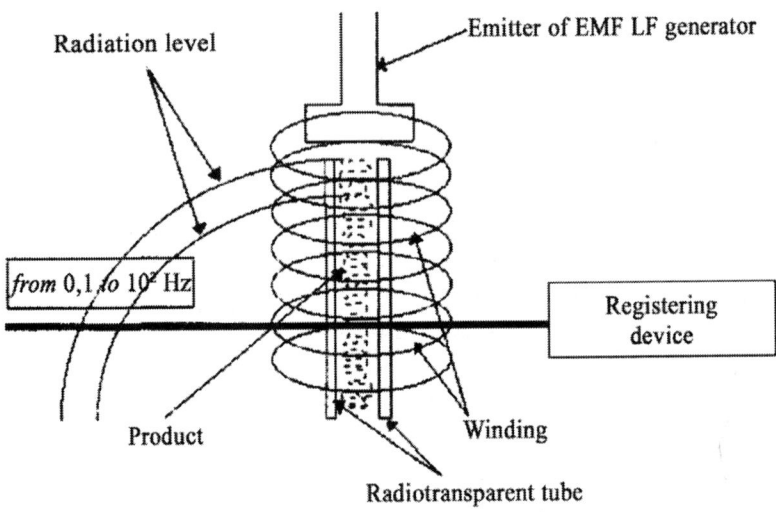

Figure 1: Scheme of processing of liquid environment by EMF LF.

EMF LF frequencies, which were used to process the fish raw materials that had to be canned, were chosen in accordance to the rapid method to determine the resonant frequencies of biological objects [1]. The simultaneous processing of the magnetic field of extremely low range and of woofer low length (field strength 1 - 150 A /m) and an alternating electric field of frequency 1 - 100 Hz (intensity 0.05 - 50 mV/m) on the researching objects changes pH length, the mass fraction of solids and the refractive index of extracts of meat raw materials.

The authors at first have established the selection effect of the resonance frequency of the objects to hundreds of Hz. Biochemical, histomorphological and microbiological evaluations of fish raw materials in biomodified under influence of EMF LF with resonance frequencies have been performed. Investigated objects have been processed during 20 - 60 min at the magnetic induction 6 mT.

Our investigations let to conclude that the impact of EMF LF (f = 40.03 Hz) on fish raw materials intensifies the maturation process of muscle tissue, also helps to change the degree of moisture connectivity, and acts as a preservation factor (barrier) against the microbiological spoilage.

The elements of barrier technology have been introduced in the industry for more effective managing by the storage of food concentrates. Low-temperature drying, the presence of CO_2-extractable complexes, gas-liquid processing and bactericidal packaging are preserving factors in the technology of fish-vegetative concentrates. The most effective barrier method is processing of raw materials and semi-finished products by EMF LF.

The authors comprehensively have investigated the effect of pre-plant seeds and post-harvest production after EMF LF-processing (potatoes, sweet pepper, tomato and cabbage) on the nutritional value, technological properties and the commercial quality of vegetables.

The effects of EMF LF processing on biological objects have been examined. The magnetic field is used to get an effect of the magnetic moments of the researched objects. Lorentz force acting on the electrically charged particle moving in a magnetic field, is always directed perpendicularly to the vectors v and B. Lorentz force is proportional to the charge of the particle g, component velocity v, perpendicular to the magnetic field vector and to the value of the magnetic field B. In the SI system of the units Lorentz force is expressed by the equation:

$$F = g[v, B],$$

(2)

where square brackets denote the vector product.

The set of electrical and magnetic fields indicator "Cyclone 04" (developed in Fryazino city, Moscow region) was used in the research. The apparatus is intended to measure RMS values of the magnetic induction and EMF LF generated by various means. The frequency range of measured signals was from 5 Hz to 2 kHz and the alternating magnetic field meter with a measuring range of 10 - 5000 nT.

RESULTS AND DISCUSSION

The method of EMF LF processing of sown seeds of plants was developed with the authors' participation. The method was produced in piles located in the closed warehouse. The activation of seeds was made by the generator of EMF EF running in the automatic mode without people's presence according to the technological requirements and coordinated with Sanitary Epidemic Department of Russian Federation. Seed volumes processing is not limited. Processing time was 4 - 6 days, depending on the specialties of agricultural raw and phytosanitary condition. One setting during the technological cycle processes up to 150 - 200 tons of seeds.

After the completion of processing, seeds are left on the ageing for the energy exchange between treated and non-treated seeds in the shoulder deep. It is explained by the fact that the EMF LF processing of vegetative raw materials tissues in the range of 6 Hz to 80 Hz results to the formation of biological resonance. Living tissues become themselves as a source of secondary radiation, which stimulates the cell division and accelerates the oxidation and the metabolism in adjacent tissues which don't experience the direct processing of electromagnetic field. The result of the secondary radiation is the energy exchange in large masses of seeds, from the surface to deeper layers of the bead. The efficiency of the between-seeds energy transfer depends on the temperature and the mass of the grain: the greater the mass of seeds, the greater the effect of the electromagnetic activation and longer the duration of its preservation (100 - 120 days). The temperature for the seed processing and binning (the aftereffect period) may be from 20°C to 30°C.

The electromagnetic activation of seed materials can be carried out in a wide range of dates, but more optimal to start processing within 20 - 25 days before sowing (planting). The effect of bio-resonance seeds and cuttings conditions disappears after about 120 days after processing. The significant cost savings compared to other methods of processing is due to the lack of supplies in the combination with the high automation of all the processes.

The estimated price of EMF LF processing of agricultural raw materials depends on the volume of products for the long term storage.

The cost of acreage processing is 0.25 cents/Ha.

The approximate profit in processing of one ton of vegetables (at the decrease of storage losses by 15%) is from 200 to 265 dollars, and of one ton of fruit and berries is from 340 to 470 dollars.

ADDITIONAL RESEARCH

The authors have investigated the intensification of technological processes with the usage of low frequency electromagnetic field.

The gas-liquid processing is one of the safe agricultural raw materials processing technologies. The liquid and compressed gases extraction as a part of this technology allows to select target components, save the negative properties of food products, apply soft thermal treatment regimes, etc.

Thus, due to the rapid development of new trends in food processing technology, the liquid and compressed gases extraction has great prospects for implementation processes of a wide range processing of agricultural raw materials and food products. Therefore the problem of the extraction intensification is decisive in implementation of such technologies.

Lots of methods to intensify the extraction process by liquid and compressed gases can be divided into three main groups. The first group includes the intensification methods through the optimization of technological parameters of the extraction process (the choice of pressure parameters, the temperature, and the preparation of raw materials). The second group includes mechanical methods of the intensification (mechanical mixing of raw materials, etc.). The third group includes the methods of influence on the physical parameters of the environment of the extraction process (processing by ultrasonic waves, infrared electromagnetic waves happens). All of the above methods of the process intensification of the extraction by liquid and compressed gases have either the direct mechanical effect or the heat effect on components of the process that is associated with a significant energy costs.

Previously, the effect of electromagnetic oscillations of non-thermal intensity on biological structures is considered insignificant, but at present the authors are working at this problem and have already received a positive effect in the processing of food and agricultural raw materials of non-thermal electromagnetic fluctuations of the intensity at low frequencies. Thus the application of such effects to intensify the process of extraction can greatly reduce the energy consumption of the extraction process.

The research of this area shows that the main problem of setting creation of intensification of the extraction process by the action of low-frequency electromagnetic waves is the creation of the precision EMF LF-generator that is stipulated by the high sensitivity of processed objects to the oscillation frequency.

The laboratory setting of intensification of the extraction process (Figure 2), that uses the developed precision low frequency sinusoidal signal generator, has been developed and it has been manufactured at Meat and Fish Products Technology Department of Kuban State Technological University.

The setting includes the precision audio oscillator of low frequency (PAOLF), the high-frequency generator (modulator—M, power amplifier—PA), the coil of inductivity (CI—acts as a radiator low-frequency electromagnetic fluctuations). The sinusoidal wave from the generator (G) is modulated by the low-frequency signal (amplitude or frequency modulation) and is converted by the CI through PA in electromagnetic waves.

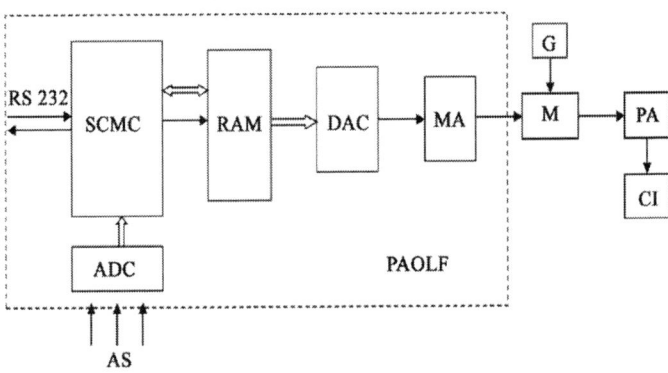

Figure 2: The laboratory setting for intensification of the extraction process.

With the purpose of development of PAOLF the principle of the digital formation of the sine wave signal of low frequency has been used: the period of a sinusoidal signal is divided into a discrete number of time quanta, each of which is formed with corresponding to a given quantum of the numerical value of the signal amplitude. PAOLF is consisted of following: the single-chip microcomputer (SCMC), the random access memory (RAM), the digital-analog converter (DAC), the analog-digital converter (ADC), the matching amplifier (MA). SCMC realizes the formation of the RAM of sinusoidal signal that is then transmitted to DAC to form an analog signal. MA provides buffering of DAC and of the noise filtration associated with the discrete amplitude DAC. ADC allows to connect analog sensors (AS) of pressure, temperature, etc. for the rapid analysis of the process parameters extraction. SCMC also has a serial interface RS-232 communication with the PC.

The developed unit provides the modes of low-frequency processing (by amplitude or by frequency) of modulated electromagnetic oscillations in the range of 1 - 100 Hz ± 0.01 Hz.

The influence effect of low-frequency electromagnetic field on the properties of wine has been studied (Figures 3 and 4) [10].

CONCLUSIONS

EMF LF processing of raw materials activates defense reactions of vegetative tissues, thus increases the ability to repair physical damages, reduces the metabolic activity, prolongs a period of profound peace that decreases losses and stabilizes the quality and the nutritional value of products during the long storage.

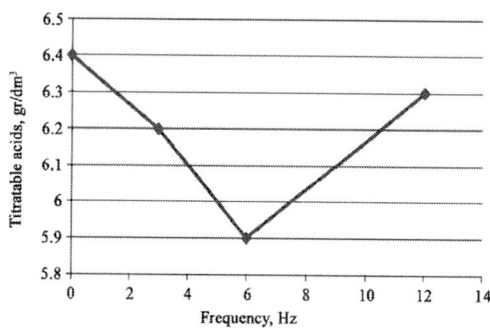

Figure 3: The influence of frequency of EMF LF processing (processing time = 45 min, B = 1.2 mT) on the change of the content of titratable acids in wine obtained from grapes of Cabernet Sauvignon.

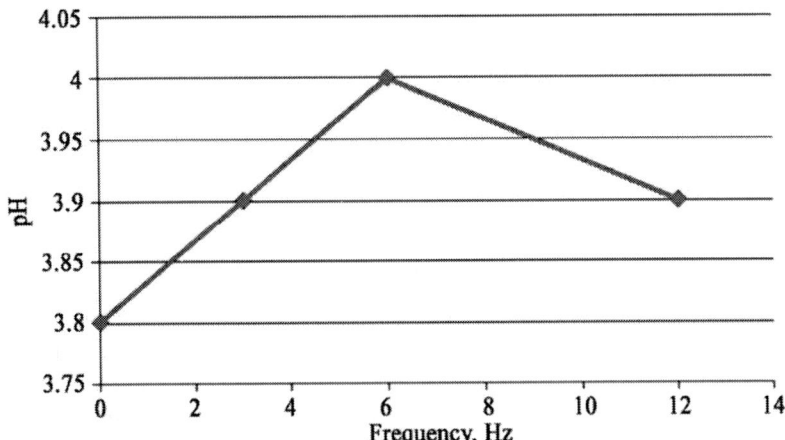

(a)

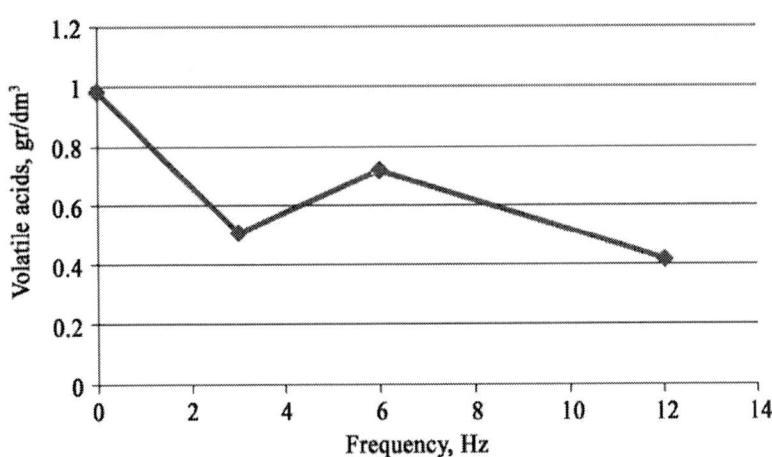

(b)

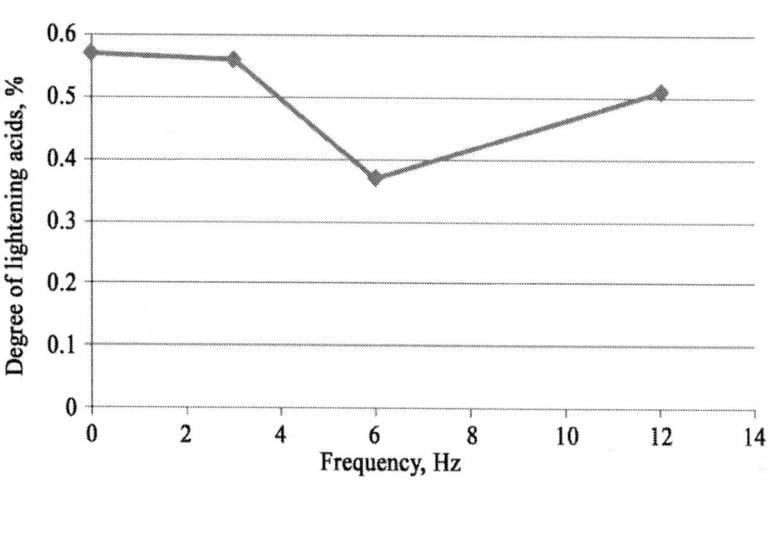

(c)

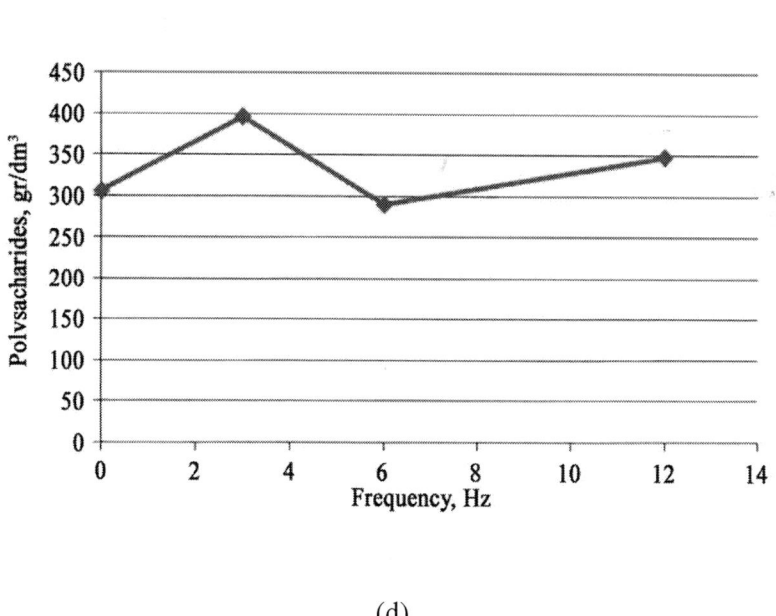

(d)

Figure 4: (a), (b), (c), (d) The influence of frequency of EMF LF processing (processing time = 45 min, B = 1.2 mT) on the change of the content of pH rate, of volatile acids, of the lightening acids degree, of polysaccharides in wine, obtained from grapes of Cabernet Sauvignon.

At first the regularities of the dependence between the change of the quality of the tomato seeds germination and the time changes of amplitude-modulated and frequency-modulated magnetic fields have been established.

For the practical usage the methodology of EMF LF processing of fish and vegetable raw materials has been proposed and applied. This methodology let to increase the quality of vegetable and fish production by the shelf life prolongation. While special attention was paid to the security of the personnel using the EMF LF generator [11].

By jury's decision of XI International Salon of Inventions and Innovation Technologies "Archimedes-2012" the development of A.V. Grachev and G.I. Kasyanov "The hardware and software signal generator of low and extremely low frequency used in the setting for electromagnetic processing of vegetative and animal raw materials origin" has won a silver medal.

REFERENCES

1. M. G. Baryshev and G. I. Kasyanov, "Electromagnetic Processing of Vegetative and Animal Raw Materials Origin," Kuban State Technological University, Krasnodar, 2002.

2. M. G. Baryshev, S. S. Djimak and G. I. Kasyanov, "Application of Water with Modified Isotopic Composition and pH in Meat Industry," Proceedings of Universities. Food Industry, No. 2-3, 2012, pp. 42-44.

3. M. G. Baryshev, G. I. Kasyanov, R. S. Reshetova and G. P. Ilychenko "The Method of Storage Sugar Beet Root," Russian Federation Patent No. 2172091, 2001.

4. M. G. Baryshev, G. I. Kasyanov, R. S. Reshetova and G. P. Ilychenko, "The Method of Storage of Sugar Beet," Russian Federation Patent No. 2172095, 2000.

5. M. G. Baryshev, G. I. Kasyanov, G. P. Ilychenko and V. V. Magerovskiy, "The Method of Processing of Seeds," Russian Federation Patent No. 2175179, 2001.

6. M. G. Baryshev and G. I. Kasyanov, "The Method of Drying of Vegetative and Animal Raw Materials Origin," Russian Federation Patent No. 2203458, 2003.

7. O. U. Gryzlova, T. I. Subbotina and A. A. Hadartsev, "Bioresonance Effects in Time of Electromagnetic Fields Impact: Physical Models and Experiment," Triada Publishing, Moscow, Tver, 2007.

8. M. G. Baryshev, N. S. Vasilyev, N. N. Kulikova and S. S. Djimak, "Low Frequency Electromagnetic Field Impact on Biological Systems," South Scientific Centre of Russian Science Academy, Rostov on Don, 2008.

9. O. U. Gryzlova, "Bioresonance Effects in Natural and Artificial Electromagnetic Fields as the Vitality Factor," Dissertation for the Degree of Candidate Technical Sciences, Tula State University, Tula, 2005.

10. V. T. Hristyuk "Improvement of Wine Technology with Application of Methods of Electrophysycal and Sorption Processing," Under Redaction of Honored Scientific Worker of Russian Federation, Honored Inventor of Russian Federation, Ecoinvest Publishing, Krasnodar, 2012.

11. A. A. Lubomudrov, "Basics of Safety in Time of Working with Sources of Electromagnetic Fields," IBT Press, Moscow, 2011.

Chapter 7

Theoretical Modeling and Experimental Analysis of Drying Process in Electromagnetic Field

Arif Memmedov[1], Teymuraz Abbasov[1],
and Mustafa Şeker[2]

[1]Department of Electric-Electronic Engineering, Faculty of Engineering, University of Inonu, Malatya, Turkey
[2]Department of Control and Automation, Divriği Nuri Demirağ Vocationally High School, University of Cumhuriyet, Sivas, Turkey

ABSTRACT

The effects of electromagnetic waves in drying processes of solid materials are investigated theoretically and experimentally. Modified model of mass transfer being constituted by the effect of electromagnetic waves which have different frequency has been obtained. Modeling of the drying process with a two-port electric circuit for the determination of diffusion coefficients is designed. The frequency limits of electromagnetic wave which will be able to hasten the drying process are determined. The effects of the electromagnetic wave in the potato slice drying process by the influence of different frequencies and temperatures are experimentally examined. The results obtained are compared with theoretical calculations. Moisture concentration curves in drying process have been commented by drawing. Theoretical and experimental results which have been obtained are identified as a well adaptation.

INTRODUCTION

Main building components of all fruit and vegetables are foreboded to be water. Per contra, microorganisms absolutely need the water for metabolisms. The life of microorganisms causes deterioration of fruit and vegetables. For this reason, the amount of water in their composition is required to reduce or totally eliminate for prolonged storage of fruit and vegetables. Prolonged protection method by means of reduction of the water contained in foods is one of the oldest methods practiced by people in food preservation. In drying of food production, dehydration process with the aid of solar energy or the heat obtained from other sources is under way. In the second method, impending "convection drying", "contact drying" and "radiation drying" according to the methods to move heat required in order to remove the water from the product dried, three drying methods are mainly used - . In addition, two speared "dielectric drying" and "discontinuous drying" methods are used in the construction and textile industry. The heat in dielectric drying, "dipoles" of the dielectric materials in high-frequency electric field composed as a result of friction obtained the displacement. The heat in discontinuous vacuum drying is obtained by bringing light of hidden energy in matter. In 1970-1980 of the last century, tobacco drying in magnetic field has been tried [5]. In this method, eddy currents occur depending on the frequency of the magnetic field and these currents cause the formation of heat. That should be taken into consideration throughout drying fruits and vegetables varying both color and chemical structure - . Optimal drying conditions in order to reduce to the lowest level of these changes should be selected. The accelerating method of classical drying systems for the creation of optimal drying conditions can be more important. In practice for this purpose, sulfur dioxide gas (SO_2), the implementation of electrical and magnetic field and blushing of fruit and vegetables [4],[6]-[9] are used for management. Magnetic and electric fields are widely used in recent years in order to accelerate the drying process of fruit, vegetables, and food materials. Under the influence of the magnetic field in drying process, events arise such as the increase of the temperature permeability of medium, temperature increasing due to the formation of electric field throughout closed-loop in the liquid medium and fragmentation of heterogeneous environments because of diversity of magnetic permeability of different phases. The effect of the electrical field in dried environment leads to the emergence of many physic-chemical phenomena. Because of different characteristics of phases dried in electrical operation, the environment of heterogeneous is performing fragmentation. In electrophoresis processes, heating of the material is performed by creating a current in the electrical field with the effect of marked electrical loads in hard and liquid phases. In addition, temperature in this medium due to electrical current passing in dried environment can be created.

In literature, study results and information belonging to drying materials in the electric field [10], [11], magnetic field [5], and electromagnetic field [8], [12] are presented. In this study, the results of experiments under the influence of constant magnetic field [5], high-frequency and high-voltage electric fields [10], [11], high-frequency (microwave) fields [8], [12], [13] and the results from the use of their different theoretical methods are presented. In other study [9], determination of the effective diffusion coefficient of humidity considering the effect of low-frequency magnetic field and deformation of dried material is determined, such as the solution of the inverse problem. In reference [14], drying porous-capillary layer of glazing spheres is taken into consideration, and the effect of drying kinetics parameters of the particles moistness and electromagnetic fields are experimentally examined. As a result of investigations, calculation of humidity transport current in porous surroundings, as the sum of resultant moving current induced by diffusion and magnetic field is proposed. All components of the theory and practice of these events including many physical and chemical changes are not possible after enough examination.

In this study, accelerating the drying processes' events effects of electromagnetic wave (EMW) of materials having different physical and chemical properties are investigated theoretically and experimentally. Experimental investigations without EMW effect and EMW effect in drying system designed especially for this purpose have been carried. EMWs having a certain frequency have been identified as accelerating the effective drying process. The creation of the theoretical model for drying effect of EMW has been attempted. Two-port circuit which is widely used and examined in detail in electric circuit theory has offered a suggestion for the determination of the basic diffusion coefficients that assess the mass transfer events in drying process. Theoretical and experimental results which have been obtained are identified as a well adaptation.

PROBLEM FORMULATION AND MATHEMATICAL MODEL

In the analysis of the electromagnetic field effect problems in mass transfer events occurred porous environment or drying process in material, provided that the fallowing conditions are assumed.

- The solid material sample which is examined composed of porous layer shaped a rectangular or circular of finite dimension in the thickness 2δ.

- Opposed surface of the porous material has heat and mass transfer by the effect of air flow and electromagnetic field. Velocity of the air which is blown off constant during the drying process and air flow is speared over homogeneously on the surface of the material.

- The effect of the electromagnetic field generated with electrical (E) and magnetic (B) fields affecting in the direction perpendicular to each other. Electromagnetic field is an alternative field.

- Heat and mass transfer formed are a one-dimensional process.

- Chemical reaction does not occur in material during the drying and the chemical structure of the material does not change. Material has a homogeneous structure.

Take into consideration these circumstances, surface of the sheet of material which dimensions L and b, thickness 2δ is exposed to a laminar air flow in the T_∞ = const initial temperature (Figure 1). M_0 is initial moisture concentration. Material is composed of capillary porous and this material undergoes changes in volume during drying. Due to the air flow that direction of y-axis, maximum transmission of air temperature will be in this direction. Instantly t = 0, electromagnetic field which have been created from electric (E) and magnetic (B) fields towards perpendicular to each other to layer has shown an impact. This field generates an electromagnetic wave (EMW) in the direction of z-axis. This EMW, in layer of material which is examined will give rise to additional heat source. Temperature created by the source will spread as a homogeneous in layer. According to the amount of moisture contained in the dried material changed during drying, the resistance exhibited to EMW of this medium or wave impedance that determines the spread in this environment of EMW will change. Heat and mass transfer are considered in mathematical model of the drying process. This process is performed by the following approach.

In general, all of the parameters that affect the process of drying of porous materials have a non-linear change according to time and temperature. Therefore a general theoretical model creation of these processes is considerably difficult. In the literature, different empirical expressions and experimental curves are presented for the parameters that determine these processes. These expressions obtained considering the local conditions are not sufficient to create a general approach to the solution of the problem. On the other hand considering some of the conditions and assumptions, simple relationships between these parameters can be obtained.

Cross-section of layer of drying material is demonstrated in Figure 2. As can be seen in Figure 2, material layer in drying process has contracted, deformed all surfaces and dimension of the layer has decreased. In other words, the maximum deformation of material layer comes into the existence in the z-axis. Considering the assumption above, the mass change in the z-axis (mass conservation) can be written as following according to the second law of Fick's equation.

$$\frac{\partial M}{\partial t} + U \frac{\partial M}{\partial t} = D_{eff} \frac{\partial^2 M}{\partial z^2} + D_{elm} \frac{\partial^2 M}{\partial z^2}$$

(1)

Here D_{elm} refers to diffusion consisting by the effect of electromagnetic field and gets the name of electromagnetic diffusion coefficient; U is shrinkage velocity of material surface. In general, heat diffusion coefficient of D_{eff} and D_{elm} have the different structure because of the dissimilar physical process of EMW events. For this reason the analytic solution of equation (1) is difficult or impossible. Thus solution of this equation can be obtained by either numerical methods or a difficult mathematical approach. In this study, equation (1) is shown as follows convective, diffusion and superposition of diffusion process comprising the impact of EMW.

$$\frac{\partial M'}{\partial t} + U \frac{\partial M'}{\partial z} = D_{eff} \frac{\partial^2 M'}{\partial z^2} \tag{2}$$

$$\frac{\partial M''}{\partial t} + U \frac{\partial M''}{\partial z} = D_{elm} \frac{\partial^2 M''}{\partial z^2} \tag{3}$$

In this place, M' resulting from convective diffusion, M'' emerging a result of diffusion consisting the effect of EMW refer to the amount of mass transfer. In this case, the amount of exact mass transfer; Determine by the expression.

$$M = M' + M'' \tag{4}$$

In general, equations of heat and mass transfer are presented numerous solutions and methods in the literature

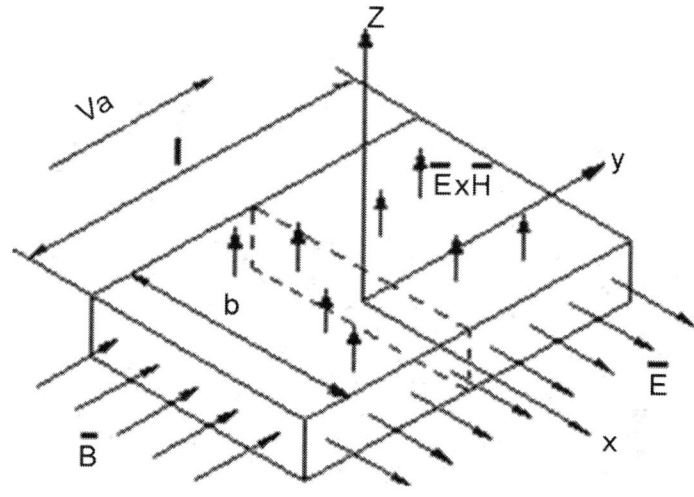

Figure 1: The general scheme of the dried sample material.

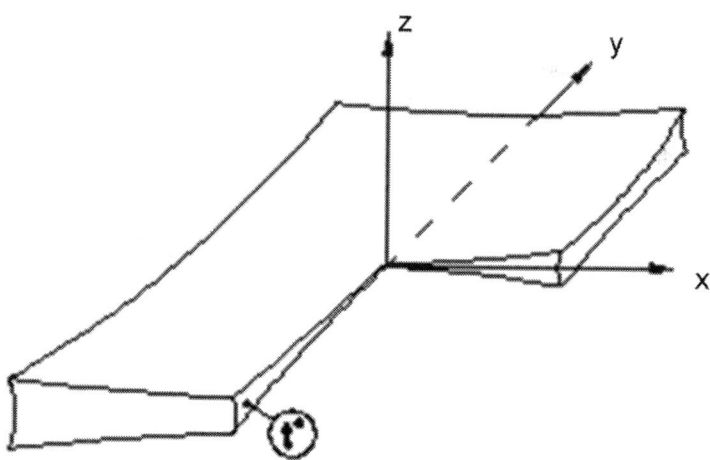

Figure 2: The geometry of dried material sample.

- , , - . Selecting the different coefficients, determination and the implementation of approximately solution methods in these solutions has been made. The results obtained consist of complex mathematical expression in many cases and are not apply handy in the practice of engineering. However taking into consideration the similarity of process flow, heat and mass transfer equations can be easily modeled with the different physical processes.

Equations expressed in many process similar to equation (2) and equation (3) used in electrical circuits and systems, and the electromagnetic theory. For example, processes expressed in similar to equation (2) and equation (3) can be modeled with the theory of two-port widely used in electrical circuit theory. The main advantages of this model are a comprehensive study of the theory of two-port in electrical circuit theory and simple analytical connections in between the parameters of the two-port.

Modeling Of Diffusion Equations with Electric Circuit Theory

Principle diagram of simple two-port system is shown in Figure 3. Events occurring in a system according to the theory of the two-port are determined corresponding relation between input (two-ports) and output (current-voltage) parameters of this system. Flow type of processes in the two-port are insignificant and the characteristic of the system is determined with the relationship between them and its input and output parameters. In generally, drying process of materials can be modeled the fallowing approach according to two-port electrical circuit theory.

Material which has dried and having a certain geometry and volume is assumed that a two-door. Input and output parameters that determine the drying process of these two-doors are as follow;

ρ—Density together with water mass of material (kg/m^3);

ρ_0—Density of liquid mass resulting volume of material by the effect of diffusion (kg/m^3);

Q—Heat current (y) entering volume of material;

Q_0—Heat current (y) emerging with the vapor of liquid from the volume of material.

In these conditions, the following equations can be written according to the theory of a two-port in accordance with the Figure 3.

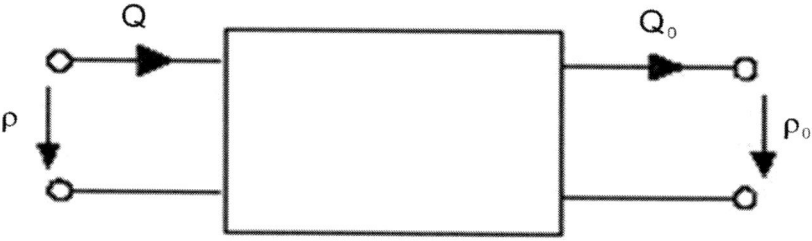

Figure 3: Diagram of a simple two-port circuit.

$$\rho = A_{11}\rho_0 + A_{12}Q_O$$

(5)

$$Q = A_{21}\rho_0 + A_{22}Q_0$$

$$A_{11} = \left(\frac{\rho}{\rho_0}\right)\Big]Q_0 = 0, \quad A_{12} = \left(\frac{\rho}{Q_0}\right)\Big]\rho_0 = 0$$

(6)

Here, A_{11}, A_{12}, A_{21}, A_{22} are the coefficient of the two-ports. These coefficients are dependent on geometry and dimension, internal structure, physical and chemical properties of the two-door or the volume of material. According to the schema of equivalent two-pole shown in Figure 3, A_{11}, A_{12}, A_{21}, A_{22} coefficients are determined as follows.

$$A_{21} = \left(\frac{Q}{\rho_0}\right)\Big]Q_0 = 0, \quad A_{22} = \left(\frac{Q}{Q_0}\right)\Big]\rho_0 = 0$$

(7)

When ω theorem applied to take into consideration the second equation in equation (5) and equation (6) and after making some mathematical operations, rate of moisture concentration can be determined as follow on convective diffusion in drying process,

$$\frac{M'}{M_0} = \frac{1}{\left[k_1 + k_2 \frac{T}{T_\infty}\left(\frac{t}{t_\infty}\right)^2 B_i\right]}$$

(8)

Here, k_1 and k_2 are constant of the system and are determined according to equation (7). In this equation, T, T_∞, t, t_∞ represent to expression temperature, the blown air temperature, time, reference time respectively and B_i is Biot number.

$$B_i = \frac{h_m L}{D_{mol}}$$

(9)

D_{mol} is molecular diffusion coefficient in equation (9) and according to Einstein formula is generally written as follows;

$$D_{mol} = \frac{kT}{m} \mu_d$$

(10)

Here, m—mass of liquid vapor separated as a result of diffusion, k—Boltzmann constant, μ_d—move capability. Mass transfer coefficient (h_m) can be determined from equation (13) as Equation (9),

$$h_m = \frac{B_i D}{L}$$

(11)

As seen in the above expression, Biot number (B_i) varies depending porous material, time and temperature. Whereas porous of material dried is not easy to determine both theoretically and experimentally. Thus, exchange porous of material in drying process can be taking into account. Due to the possibility of measuring time and temperature during the experiment,

$$\alpha(t) = B_i \left(\frac{T}{T_\infty}, \frac{t}{t_\infty} \right)$$

(12)

Making the transformation, equation (8) can be written the simplest possible manner.

$$\frac{M'}{M_0} = \left[k_1 + k_2 \alpha(t) \right]^{-1}$$

(13)

Therefore, taking advantage of a two-port theory (equation (5) and equation (6)), moisture proportion of convection drying process according to equation (13) can easily calculate by determining k_1 and k_2 coefficients taking into account physical and chemical properties, dried material structure. In this approach, a(t) is a function that determines the variation of drying process in the material and time-varying. This function is identifying by empirical method depending on the character of drying process. The effect of EMW in drying process can achieve the fallowing approach.

Analysis of the Effect of EMW's Drying Process

If we have taking into account that water is essentially chemical composition of the dried materials, we can assumed that these materials are essentially non-magnetic or very weak magnetic (diamagnetic). Hence, magnetic permeability of the dried material becomes $\mu = \mu_r$ $\mu_0 = \mu_0 = 4\varpi \times 10^{-7}$ H/m. The other hand, since they have a certain electrical conductivity, dried material can be assumed that a partially conductive ambient. In this instance, Maxwell's equations are written as follows by assuming $\varepsilon \geq 100$ [5-9].

$$\nabla x H = \left(\sigma + jw\epsilon\right) E$$

(14)

$$\nabla x E = -jw\mu H$$

(15)

$$\nabla x E = 0, \nabla x H = 0$$

(16)

In the event of (z-dimension) one-dimension propagation of electromagnetic waves, charges in fields of E and H in these equations are obtained as follows;

$$E = E_0 e^{-yz} a_x$$

(17)

$$H = \sqrt{\frac{\sigma + jw\epsilon}{jwM}} E_0 e^{-yz} a_y$$

(18)

EMW's wave impedance ZD sprawling in the z direction in dried ambient and diffusion coefficient γ,

$$Z_D = \sqrt{\frac{jw\mu}{\sigma + jw\varepsilon}}$$

(19)

$$\gamma = \alpha + j\beta$$

(20)

as are determined. Here, α is wave extinction coefficient and β is phase constant. If these expressions taken into consideration, solution of the wave equation (equation (17) and equation (18)) is written as follows .

$$E(z,t) = E_0 e^{-\alpha z} e^{j(wt - \beta z)} a_x$$

(21)

$$H(z,t) = \frac{E_0}{[Z_D]} e^{-\alpha z} e^{j(wt - \beta z - \theta)} a_y$$

(22)

Here,

$$[Z_D] = \frac{\sqrt{\frac{\mu}{\epsilon}}}{\sqrt[4]{1 + \left(\frac{\sigma}{\omega\varepsilon}\right)^2}}, \quad \tan\theta = \frac{\sigma}{\omega\epsilon}$$

(23)

Propagation velocity of the electromagnetic wave and wave length is calculated as follows.

$$U = \frac{\omega}{B}, \lambda = \frac{2\pi}{\beta}$$

(24)

According to wave parameters contained in equation (21) is dependent on the physical characteristics of dried material, these parameters will have different values in distinct process. For this reason, the corrected values of these parameters should be determined according to the measurement taken from drying experiments. For this purpose, drying power formed by the electromagnetic wave must be measured. This process can be done using a special cell designed in drying mechanism (Figure 4).

In this drying cell given in description of the next section can readily be determined change over time the electrical conductivity of

the dried material. According to these results, the proportion of ($\frac{\sigma}{\omega E}$) in dried system can be evaluated and cut-off frequency f_c required by drying system can be determined. Cut-off frequency and physical properties of the dried material, blow out and phase constant can be easily calculated,

$$\alpha = \omega \sqrt{\frac{\mu}{\varepsilon} \left[\sqrt{1 + \left(\frac{\sigma}{\omega \varepsilon} \right)^2} - 1 \right]}$$

(25)

$$\beta = \omega \sqrt{\frac{\mu}{\varepsilon} \left[\sqrt{1 + \left(\frac{\sigma}{\omega \varepsilon} \right)^2} + 1 \right]}$$

(26)

After calculating these parameters, changes in areas of E and H from equation (12) can be easily determined. After determination of the E and H fields, the contribution of EMW in drying process and proportion of the change in moisture by the effect of this contribution can be calculated. The contribution of the heat shown the drying process of EMW can be evaluated by Poynting vector.

$$P = \left[ExH^* \right]$$

(27)

Calculated from the real component of this vector producing the effects of thermal drying process is part of the active power.

$$P_a = \frac{1}{2} R_e \left[ExH^* \right]$$

(28)

Here, H* is the complex conjugate of the H magnetic field strength. Heat energy generated the dried materials of EMW,

$$W = \frac{t_1}{t_2 - t_1} \int_{t_1}^{t_2} P \mathrm{d}t = \frac{t_1}{t_2 - t_1} \int_{t_1}^{t_2} R_e \left[ExH^* \right] \mathrm{d}t$$

(29)

as determined. Where E and H* fields are determined according to the calculated values in Equation (21).

Heat generated by the induction in dried material,

$$P_a = \sigma f_c^2 B_m^2$$

It can be calculated according to the power. Wherein B is the magnetic field density and is calculated accord

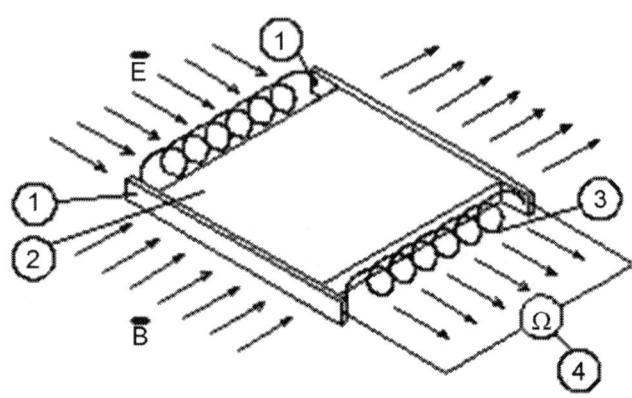

Figure 4: Block diagram of the cell designed for the determination of drying power.

ing to equation (29). Taking into consideration parameters of EMW, proportion of the change in moisture by the effect of EMW using the ϖ theorem can be modeled as follows,

$$\frac{M^{''}}{M_0} = \left[120 \left(\frac{f}{f'} \right)^2 \left(\frac{t}{t_\infty} \right) \alpha_2 (t) \right]^{-1}$$

(30)

Here f = 100 kHz is cut-off frequency of wave. Thus take into consideration Equation (13) and Equation (30), variation of total moisture concentration in drying process is obtained in the following format from M in (4).

$$\frac{M}{M_0} = \frac{M^{'}}{M_0} = \frac{M^{''}}{M_0} = \left[k_1 + k_2 \alpha (t) \right]^{-1} + \left[120 \left(\frac{f}{f'} \right)^2 \left(\frac{t}{t_\infty} \right) \alpha_2 (t) \right]^{-1}$$

(31)

EXPERIMENTS

Examine the effect of drying process of electromagnetic wave has been made in laboratory drying machine specially designed [9]. The main elements of drying cell and drying system has been shown in Figure 5.

Material sample 4 dried in drying regions which has cylindrical form has been placed. Drying pipe 1 has been made from insulator and non-magnetic material (plastic). Electrodes of 2 have been placed in the inner walls of the pipe. The geometry of electrodes and dimensions has been selected so that; electric field E that these electrodes are formed get a homogeneous spread on the dried material. The magnetic field (B) in drying region is created by solenoid 3 wrapped on the drying pipe. Dried sample and solenoid 3 according to dimensions of the drying system be considered as a long solenoid. Therefore, B field formed in drying pipe of solenoid 3 has been evaluated as homogenous area. The effect of electromagnetic field on the dried sample 4 has been implemented as electric (E) and magnetic fields (B) that direction perpendicular to each other due to the structure of 2's electrodes and 3's solenoid. The air which have capable maintaining constant temperature (T = constant) within the drying pipe during drying has been blow as a laminar at a constant speed (V = constant). Dried sample 4 which have dimensions of $2\delta \times b \times L$ are placed on the center axis x of drying pipe and right in the middle. Thus, the dried sample 4 remains under the influence of E and B fields of air blown continuously during the drying. E and B fields and acting on EMW can be controlled and adjusted by the external electric circuit. The geometry of sample 4 dried periodically during drying is examined and mass change is measured.

The power of drying system to assess the effect of EMW is determined by measurement of the drying cell shown in figure 4. For this purpose, sample dried in drying process is measured resistance and permeability. Drying cell in figure 3 has been created that two electrodes thickness of 1 mm, ohmmeter which is connected to copper conductor wire of these electrodes. Copper electrodes have been combined with springs to provide the flexibility of drying cell volume. Dried sample 2 is placed between the copper electrodes. Varying resistance and permeability with volume change as a result of drying process of dried material can be continuously measured by ohmmeter 4.

RESULT AND DISCUSSION

Drying of different materials have been made for assess the impact of the drying events of electromagnetic wave and comparisons of the results obtained from theoretical models. The role of the effect of EMW in drying process has been determined by changing the temperature of the environment. Potato slices as the dried material sample has been used. Potato slice which have dimensions (1 mm × 3.6 mm × 5.2 mm) is placed in drying pipe 1 as shown infigure 5. The drying process has been made both under the influence of EMW and without EMW. The parameters of external search which are constitute of EMW have been taken U = 30 V, I = 0.275 A, f = 30 kHz. Rate and temperature of the blown air have been V_a = 2 m/s, T = 40°C respectively. Resistance or permeability change of sample that is periodically dried during the drying has been measured by the cell schema in figure 3 and the results obtained are provided inTable 1.

Cut-off frequency of EMW in drying process,

$$\frac{\sigma}{\omega \epsilon} \leq \approx \frac{1}{3}$$

(32)

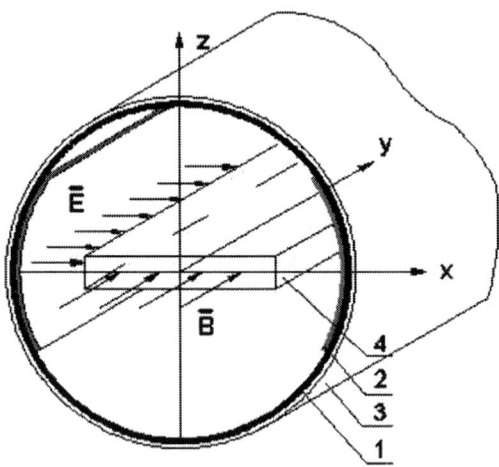

Figure 5: Principle diagram of drying mechanism; 1—non-magnetic cylindrical drying pipe, 2—electrodes, 3—swathe-solenoid, 4—dried material sample.

Table 1: Measurement of resistance and permeability changes in drying of potato samples

No	t, h	R, kΩ	, $(\Omega m)^{-1}$
1	0	73.5	1.96×10^{-3}
2	1	85	1.69×10^{-3}
3	2	107	1.35×10^{-3}
4	3	134	1.075×10^{-3}
5	4	313	0.46×10^{-3}
6	4.5	15×10^3	9.6×10^{-6}
7	5	24×10^3	6×10^{-6}
8	5.5	30×10^3	4.8×10^{-6}

Taking into consideration the inequality, cut-off frequency determined as,

$$f_c = \frac{\omega}{2\pi} \geq \frac{3\sigma}{2\pi\varepsilon} \tag{33}$$

Cut-off frequency values stemming from the change permeability of the dried material with time has been shown in Table 2. Here, for a slice of wet potato, $\varepsilon_r = 140$, $\varepsilon_0 = 8.854 \times 10^{-12}$ F/m, $\varepsilon = \varepsilon_r \varepsilon_0 = 1.24 \times 10^{-9}$ F/m, $\mu = \mu_0 = 4\varpi \times 10^{-7}$ H/m are assumed.

The effect of frequencies over f = 17 Hz after 4 hours of drying process, the effect of frequencies over f = 4 kHz after than 4.5 hours become zero. In other words, potato slice dried under the above conditions indicate a very weak conductivity in order to EMW in frequencies above approximately f = 17 kHz. In this event, the effects on drying process of EMW enough to be neglected weak. According to permeability of dried material is relatively high at the inception of drying material, EMW varying with frequency have an active part in drying process. Thought drying process is formation by the influence of both wave frequency and heat, wave frequency affecting the drying process may be selected around f = 15 kHz or less than f = 17 kHz.

Due to the E and H fields are E = 300 V/m, H = 200 A/m in the absence of the dried sample in the drying pipe at 313 K, characteristic impedance or wave impedance of the media is become as fallow,

$$Z_D = Z_0 = \frac{E}{H} = 1.5\Omega$$

(34)

Change over time of wave impedance of potato slice such as the default partially the conducting medium calculated making use of Table 1 are shown in Table 3.

In consideration of these changes, extinction and phase coefficients of the EMW calculated according to min (29) are presented in Table 4. As shown in table 2 after the time t = 4.5 hours, extinction coefficient (α) reaches the negligible lower values. This result means that the resistance to the effect of EMW-dried material.

Change over time of moisture content obtained from drying the potato slice in figure 6 has been shown. In figure 6, the absence of action by EMW in the first curve (f = 0) and the effect of EMW in drying process in the second curve (f = 5 kHz) have been shown. a(t) coefficients which is characterizing the change with time of dive .Fusion coefficient in these process have been modeled as follows;

Table 2: Cut-off frequency changes over with the time in drying of potato slice

t, h	s^{-1}
0	0.75×10^6
1	0.65×10^6
2	0.52×10^6
3	41.5×10^6
4	17.7×10^6
4.5	3.7×10^6
5	2.3×10^6
5.5	1.85×10^6

Table 3: Change over time of wave impedance

t, h	α, Np/m	β, rad/m
0	0.0145	0.93
1	0.0133	0.875

2	9.5×10^{-3}	0.694
3	8.8×10^{-3}	0.66
4	5.76×10^{-3}	0.43
4.5	1.54×10^{-4}	0.429
5	9.6×10^{-5}	0.429

Table 4: Change over time of extinction and phase constant of EMW

t, h	Z_D, Ω
0	$11.03, cis41.6^0$
1	$11.85, cis41.1^0$
2	$15.58, cis38.18^0$
3	$16.5, cis37.31^0$
4	$31.96, cis4.4^0$
4.5	$32.1, cis1.18^0$
5	$32.1, cis0.74^0$

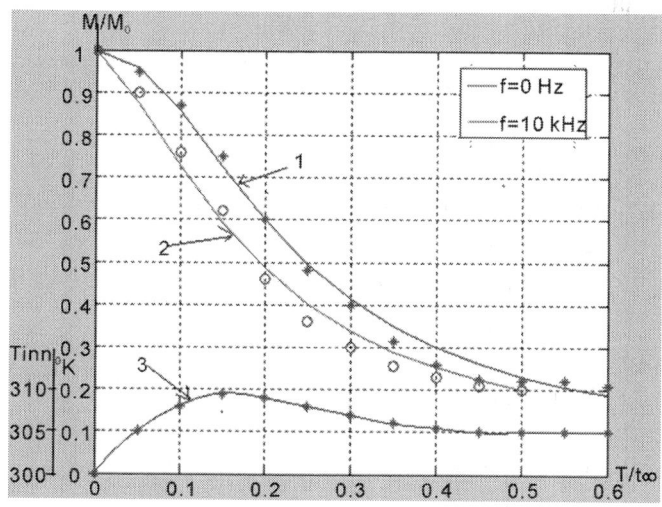

Figure 6: Moisture and inner temperature profiles of potato during drying (T = 313 K°, V_a = 2 m/s) (1—without EMW, 2—with EMW, 3—

change of the inner temperature in drying sample; continuous lines are theoretical and symbols are experimental results).

$$\alpha_1\left(t\right)=2.818e^{-\left[0.5+\left(\frac{t}{t_\infty}\right)^2\right]}$$

(35)

$$\alpha\left(t\right)=1.749\times2^{-2\left(\frac{t}{t_\infty}\right)^2}$$

(36)

$_1$(t) and (t) at issue in equation (35) and equation (36) are expressions that takes into account changes with over time of diffusion coefficient. $_1$(t) is diffusion coefficient of convection drying and α(t) is considers both the effect of convection and the effect of electromagnetic fields. These parameters are varying depending on nonlinear with time and temperature. In generally, determination of these parameters is virtually impossible in theory. Therefore, these parameters are experimentally obtained from the analysis of experimental data as dependently on the character of the drying process.

As shown in figure 6, drying process began to develop more rapidly from the first moment the effect of EMW. For example, although for the reduction %50 of humidity ratio in classical drying process need t = 5 h, this period time has been shortened as t = 4 h by the application of EMW. Therefore, drying process of potato slice by the effect of EMW has been effectively accelerated. As shown in figure 6, the results obtained from experiments in drying process provide good agreement with theoretical model presented in this study. This result also shows that the presented theoretical model can be used to model in the drying process. The curve which is number three is shown the heat exchange in this example in the drying process of potato slice.

In figure 7, the results of drying process without the influence and influence of the EMW at T = 313 K of potato slice are provided. The parameters of EMW have been selected the same as the ones used in figure 6. But the wave frequency have been adjusted as f = 10 kHz. As shown in figure, moisture content of potato slice has been decreased more rapidly with increasing temperature and drying process has decelerated. The effect of EMW on drying process under the same conditions remains approximately the same. Despite this the effect of

EMW on drying process with increasing frequency has provided to be faster of this process. Although 4 hours needed to decrease fifty percent according to the first position of humidity ratio in classic drying process, this process with the effect of EMW which $f = 10$ Hz has been around $t = 3.2$ hours.

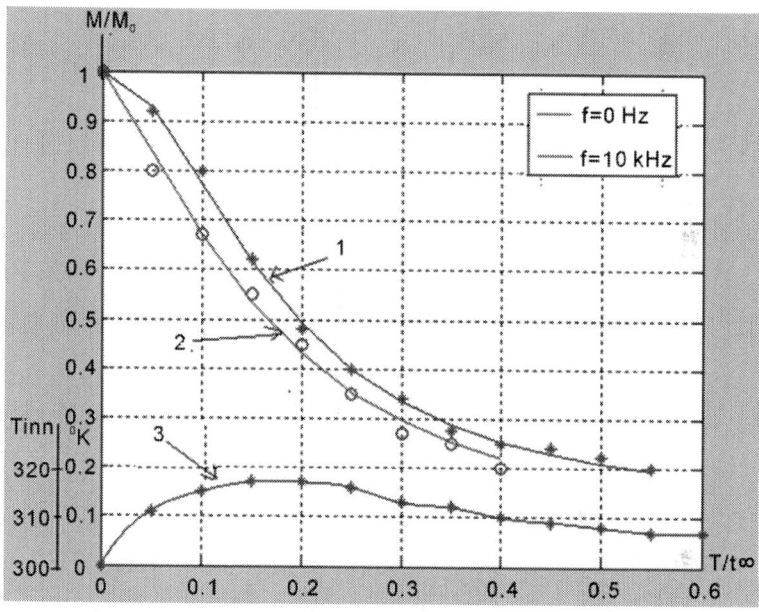

Figure 7: Moisture and inner temperature profiles of potato during drying ($T = 323$ K°, $V_a = 2$ m/s) (1—without EMW, 2—with EMW, 3—change of the inner temperature in drying sample; continuous lines are theoretical and symbols are experimental results).

The results of the drying process with the effect of EMW and without the influence of EMW of potato slice has been presented in figure 8. As a result of the increase in the frequency of EMW along with increasing of heat, drying process has been accelerated. Although it should at $t = 3.4$ h by the classical method for the reduction of %50 moisture rate, this process is reduced as $t = 2.5$ h by the effect of EMW having $f = 100$ kHz frequency.

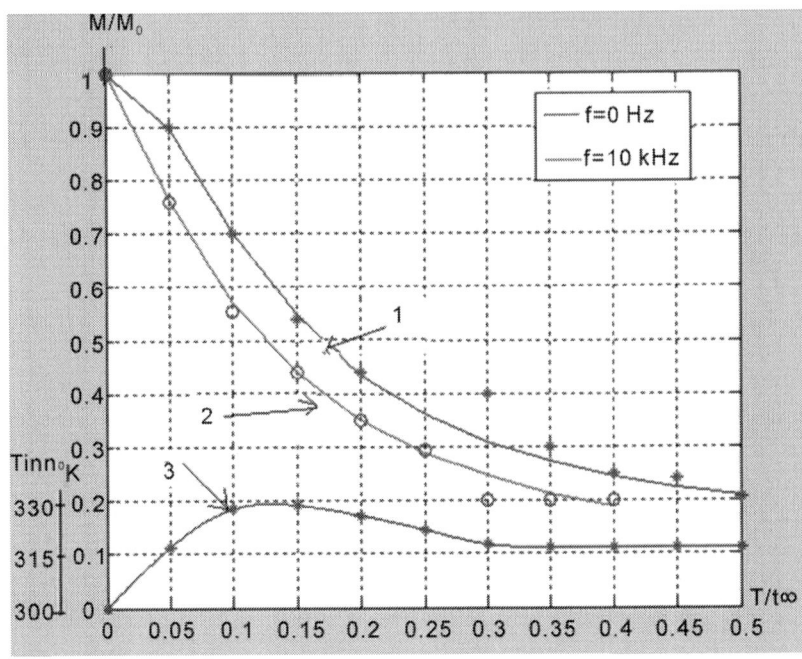

Figure 8: Moisture and inner temperature profiles of potato during drying (T = 333 K°, V$_a$=2 m/s) (1—without EMW, 2—with EMW, 3—change of the inner temperature in drying sample; continuous lines are theoretical and symbols are experimental results).

In other words, drying time in all conditions by the effect of EMW has been visibly decreased. This result proves that advantage and an effective role in drying process of EMW. On the other hand, exchange 20-fold of drying frequency by the effect of EMW have provided about 1.8-fold reduction in drying process. Therefore, in drying process by the effect of EMW as predicted expectations according to results in theoretical reviewers, wave frequency is located in an effective factor. In case of further raise the frequency of the EMW (MHz-GHz at around), conversion of microwave heating process rather than drying of this process will be caused.

CONCLUSIONS

The theoretical and experimental investigations of the effect of the electromagnetic wave in the solid material drying have been able to arrive at the following conclusions.

- The effect of electromagnetic fields in drying process remains as an important issue of the agenda and the results of theoretical and experimental investigations are presented in the literature. Microwave electromagnetic fields are influenced by the structure of substance, so the use of this model is limited in drying process of the food. For this reason, use of the weak and medium-frequency of electromagnetic waves is foreseen to accelerate the drying process of porous food materials.

- Experimental investigations of heat transfer in the dried material allow obtaining approximately the expression of mass conservation equation for accelerating the drying process under the influence of electromagnetic wave. The equation obtained from correction of empirical coefficients contained in this equation according to characteristic of the drying process and the physical structure of the material can be used for drying process for many different food materials.

- The use of model and concepts well researched in various areas of science to examine the effect of electromagnetic waves in drying process can provide a lot of advantages. The use of electrical circuit theory and π theorem enables to easily model the event of heat and mass transfer in drying process.

- According to the results of theoretical and experimental investigations, the frequency of electromagnetic wave which is accelerating the drying process of dried material is found to be 10 - 11 kHz and low frequencies. The effects of electromagnetic waves which have $f = 20$ kHz and higher values at low frequencies have been observed to be effective in the first stage of drying process.

- Theoretical and experimental results which are obtained have shown a minor disagreement. The cause of these errors occurs from being the different structure and physical properties of dried material. Therefore, the theoretical model which is presented allows taking into account a good enough approach to the effect of electromagnetic waves in drying process.

- The mechanism of action of the drying process with the electromagnetic field and waves in experiments is foreseen to complete removal of moisture rate in the material in order to evaluate all aspects. Drying the certain humidity of the material in industry environment is required. The graphs which are obtained allow determining the moisture content of material at the different stages of the dryinfg process having regards to the initial mass of dried material. For this purpose, it is useful to be preferred as the diffusion coefficient is time-varying function.

- Theoretical and experimental results which are presented in accordance with a good adaptation of the model taking into account the drying conditions also contemplate the use and considerations for drying of a broad spectrum of foods.

REFERENCES

1.   Cemeroglu, B. and Acar, J. (1986) Fruit and vegetable processing technology. Ankara.

2.   Arthur, J.C. and McLe More, T.A. (1955) Sweet potato dehydration. Effect of processing conditions and variety of dehydrated products. Journal of Agricultural Food Chemistry, 3, 782-787.http://dx.doi.org/10.1021/jf60055a007

3.   Cording, J., Willard, M., Eskew, R.K. and Sullivan, J.F. (1957) Advanced in the dehydration of mashed potatoes. Food Technology, 11, 236-240.

4.   Molatson, L.J., Spadaro, L.J., Roby, M.T. and Lee, F.H. (1962) Dehydrated diced sweet potatoes a pilot plant process and product evaluation. Food Technology, 16, 101-104.

5.   Aladjadjyan, A. and Vlieva, T. (2003) Influence of stationary magnetic field on the early of the development of tobacco seeds. Journal of European Agriculture, 4, 131-137.

6.   Tayaraman, K. and Gupta, D.K. (2005) Drying of fruits and vegetables. In: Handbook of Industrial Drying, Merket-Dekker, New York, 643-690.

7. Mc Minn, W.A. and Magee, T.R. (1997) Physical characteristics of dehydrated potatoes. Journal of Food Engineering, 33, 37-48. http://dx.doi.org/10.1016/S0260-8774(97)00039-3

8. Vilkov, G.A. and Zabelina, T.N. (1996) Diffusion in a porous system in grossed electric and magnetic fields. Journal of engineering Physics and Thermophysics, 31, 1295-1300.

9. Kelbaliyev, G.I. and Memmedov, A. (2009) The time depending distribution function and sedimentation properties of dispersion particles. Journal of Dispersion Science and Technology, 30, 1073-1078. http://dx.doi.org/10.1080/01932690802598762

10. Sadek, S.E., Fax, R.G. and Hurwitz, M. (1972) The influence of electrical fields on convective heat and mass transfer from a horizontal surface under forced convection. Journal of Heat Transfer, 94, 144-148. http://dx.doi.org/10.1115/1.3449885

11. Knorr, D. and Angebach, A. (1998) Impact of high electric field pulses on plant membrane permecabilisation. Trend in Food Science and Technology, 9, 185-191.http://dx.doi.org/10.1016/S0924-2244(98)00040-5

12. Jezek, D., Tripalo, B., Karlovic, D., Vikic-Topic, D. and Herceg, Z. (2006) Modeling of convective carrot drying. Croatuca Chemica Acta, 79, 385-395.

13. Karim, M.A. and Hawlader, M.N.A. (2005) Drying characteristics of banana: Theoretical modeling and experimental validations. Journal of Food Engineering, 70, 35-45.http://dx.doi.org/10.1016/j.jfoodeng.2004.09.010

14. Rotanadechoab, P., Aokib, K. and Akahoreb, M.A. (2001) Numerical and experimental study of microwave drying using a rectangular waveguide. Drying Technology, 19, 2209-2234. http://dx.doi.org/10.1081/DRT-100107495

15. Fortes, M. and Okos, M.R. (1980) Drying theories: Their bases and limitations as applied to foods and grains. In: Mujumder, A.S., Ed., Advances in Drying, 1, Hemisphere, Washington DC, 119-154.

16. Luikov, A.V. (1975) Systems of differential equations of heat and mass transfer in capillary-porous bodies (review). International Journal of Heat and Mass Transfer, 18, 1-14.http://dx.doi.org/10.1016/0017-9310(75)90002-2

17.   Rossen, T.L. and Hayakawa, K. (1977) Simultaneous heat and mass transfer in dehydrated food: A review of theoretical models. AICHE Symposium, 73, 71-86.

Chapter 8

Compound Auroral Micromorphology: Ground-based High-speed Imaging

Ryuho Kataoka[1, 2], Yoko Fukuda[3], Yoshizumi Miyoshi[4], Hiroko Miyahara[5], Satoru Itoya[6], Yusuke Ebihara[7], Donald Hampton[8], Hanna Dahlgren[10, 9], Daniel Whiter[11], and Nickolay Ivchenko[9]

[1]National Institute of Polar Research, 10-3 Midori-cho, Tachikawa 190-8518, Tokyo, Japan

[2]Department of Polar Science, The Graduate University for Advanced Studies (SOKENDAI), 10-3 Midori-cho, Tachikawa 190-8518, Tokyo, Japan

[3]Department of Earth and Planetary Science, University of Tokyo, 7-3-1 Hongo, Bunkyo-ku, Tokyo 113-0003, Japan

[4]Solar-Terrestrial Environment Laboratory, Nagoya University, Furo-cho, Chikusa-ku, Nagoya 464-8601, Japan

[5]Musashino Art University, 1-736 Ogawa-cho, Kodaira-shi, Tokyo 187-8505, Japan

[6]Japan Science Foundation, 2-1 Kitanomaru Park, Chiyoda-ku, Tokyo 102-0091, Japan

[7]Research Institute for Sustainable Humanosphere (RISH), Kyoto University, Gokasho, Uji, Kyoto 611-0011, Japan

[8]Geophysical Institute, University of Alaska Fairbanks, 903 Koyukuk Drive, Fairbanks 99775-7320, AK, USA

[9]School of Electrical Engineering, KTH Royal Institute of Technology, Stockholm, S10044, Sweden

[10]School of Physics and Astronomy, University of Southampton, Southampton, UK

[11]Finnish Meteorological Institute, Helsinki, Finland

ABSTRACT

Auroral microphysics still remains partly unexplored. Cutting-edge ground-based optical observations using scientific complementary metal-oxide semiconductor (sCMOS) cameras recently enabled us to observe the fine-scale morphology of bright aurora at magnetic zenith for a variety of rapidly varying features for long uninterrupted periods. We report two interesting examples of combinations of fine-scale rapidly varying auroral features as observed by the sCMOS cameras installed at Poker Flat Research Range (PFRR), Alaska, in February 2014. The first example shows that flickering rays and pulsating modulation simultaneously appeared at the middle of a surge in the pre-midnight sector. The second example shows localized flickering aurora associated with growing eddies at the poleward edge of an arc in the midnight sector.

FINDINGS

Introduction

In the low-beta plasma condition of the magnetosphere-ionosphere (M-I) coupled region, Alfvén waves hold time varying electric field parallel to the ambient magnetic field (Hasegawa [1976]; Goertz and

Boswell [1979]). This is a consequence of the electron inertia when the wavelength perpendicular to the ambient magnetic field is an order of electron skin depth. Since the spatial scale is typically approximately 1 km when mapped to the ionosphere, the inertial Alfvén waves (IAWs) are one of the most important theoretical bases to understand the diverse microstructures of rapidly varying aurora. Observations of fine-scale auroral morphology are therefore important to visualize the fundamental wave-particle interactions working in the M-I coupled region, which are also potentially useful to diagnose the local plasma environment.

Flickering aurora consists of small-scale columns (1 to 12 km width and 10 to 40 km height) with periodic intensity variations (3 to 15 Hz) in discrete auroral arcs and is often observed associated with auroral breakup events (Kunitake and Oguti [1984]). Temerin et al. ([1986], [1993]) suggested that electromagnetic ion cyclotron (EMIC) waves that occur below the inverted-V acceleration region can accelerate and modulate the field-aligned electrons over a broad energy range to produce flickering aurora. The EMIC wave model is similar to the IAW model except that it includes finite frequency effects. A possible source of the EMIC waves is instability in the double layer or electrostatic shock that produces the electron beam. The appearance of flickering aurora can be modeled as the interference of IAW or EMIC waves (Sakanoi et al. [2005]). Gustavsson et al. ([2008]) showed that a variety of structures of flickering aurora can be modeled by varying the parameters of multiple interfering EMIC waves. Whiter et al. ([2008]) showed that flickering aurora is linked temporally to auroral activity, but not spatially on small scales, which is consistent with the interfering EMIC waves. Whiter et al. ([2010]) showed that the parallel phase velocity of the EMIC waves is the primary factor in determining the energy of wave-accelerated electrons responsible for flickering aurora, which is consistent with the resonant acceleration and deceleration model of IAW/EMIC waves in combination with inverted-V acceleration as suggested by Chen et al. ([2005]). Yaegashi et al. ([2011]) and Kataoka et al. ([2011b]) showed several examples that the observed spatiotemporal scale of monochromatic flickering events is consistent with the O+ EMIC waves. Michell et al. ([2012]) showed that there is a lack of flickering spectral power at perpendicular wavenumber of larger than 2×10^{-3} m^{-1}, suggesting the existence of a minimum spatial scale for flickering auroral patches of approximately the O+ ion gyro-radius (approximately 1 km) at the interaction altitude. The expected altitude of modulation source (assuming the O+ EMIC waves) varies widely from 2,000 km to 7,000 km (e.g., Sakanoi et al. [2005]; Whiter et al. [2010]; Yaegashi et al. [2011]).

It has been suggested for a long time that dynamic aurora structures are energetic. Hallinan ([1976]) suggested that 'spirals' of >50 km imply upward current, and the threshold current density for distortion of an auroral arc was estimated as $2.5 \times 10^{-6}\,A\,m^{-2}$. Electrostatic shear-induced instability of a thin current sheet (Wagner et al. [1983]) can explain the formation of auroral 'folds' (10 to 50 km) and 'curls' (approximately 5 km) (Hallinan and Davis [1970]). The threshold current for the double-layer formation associated with curls and folds may also be assumed as the same order of magnitude as required for spirals. Ivchenko et al. ([2005]) showed that curls are caused by the precipitation of energetic electrons with a lack of low-energy precipitation, while in the 'rays' both high- and low-energy precipitation were present simultaneously, suggesting that curls are caused by the electrostatic instability of the precipitating electron sheet, while rays are likely to be a result of IAW. Chaston and Seki ([2010]) suggested that the formation of curls and folds may indicate the existence of a resistive layer which may be considered as the auroral acceleration region, while absence of resistive layer may be consistent with the recently found rapidly varying boundary features known as 'ruffs' (Dahlgren et al. [2010]). The smallest-size distortions of auroral arcs are the 'filaments' of approximately 100 m width, and the rapidly varying very narrow auroral elements are associated with the precipitation of mono-energetic electrons (Lanchester et al. [1997]). It is important to note that curls and filaments are more energetic or associated with higher electron fluxes than surrounding aurora structures (Lanchester et al. [2009]). Dahlgren et al. ([2008a]) reported that some auroral filaments are caused by higher energy precipitation within regions of lower energy precipitation, whereas other filaments are the results of a higher flux compared to the surroundings. It was also found that high-energy precipitation corresponds to discrete and dynamic features, including curls, and low-energy precipitation corresponds to auroral signatures that were dominated by rays (Dahlgren et al. [2008b]). Further, Dahlgren et al. ([2012]) identified that mono-energetic approximately 8 keV electron precipitations caused extremely narrow (approximately 70 m) and highly dynamic auroral filaments.

Some other fine-scale rapidly varying auroras were reviewed by Sandahl et al. ([2008]). However, it has been rare that both dynamic spatial variations, such as curls and folds, and rapid time variation of flickering are discussed together, although both features are complementary to diagnose the plasma environment of the M-I coupled region. Also, to the authors' knowledge, there are no examples showing the coexistence of flickering aurora and pulsating aurora especially with its fast '3 ± 1 Hz' modulations (Royrvik and Davis [1977]) at the same time. Note that flickering and pulsating auroras are the results of two completely separate particle precipitation mechanisms. Flickering aurora is associated with particle acceleration of IAW/EMIC waves, whereas pulsations are associated with pitch-angle scattering of chorus waves. Also, flickering aurora is usually associated with breakup in the evening hours, whereas pulsations are more common in the post-midnight early morning. In this letter, we report two interesting examples of fine-scale compound auroral structures and interpret the simultaneous observations of the auroral features with respect to the known morphological categories to understand the source mechanisms acting concurrently in the magnetosphere-ionosphere coupled system.

Instrumentation

In February 2014, a new high-speed camera system (NIPR-CMOS) was installed at Poker Flat Research Range (PFRR), Fairbanks, Alaska, and was operated until April 2014. The magnetic latitude is 65.7° at PFRR, and magnetic midnight is approximately at 11:30 UT. The Hamamatsu scientific complementary metal-oxide semiconductor (sCMOS) camera (ORCA-Flash 4.0, Hamamatsu Photonics, Hamamatsu, Japan) is equipped with NIKKOR 50 mm F1.2 lens (Nikon, Tokyo, Japan) without an optical filter. While the fastest sampling rate of 100 frames per second (fps) is possible with the original pixel array size of 2,048 by 2,048, we applied 4 by 4 binning and an exposure time of 0.02 s (50 fps) to enhance the photon count. The camera system was designed to obtain significant counts for bright aurora of >10 kR at green line. The field-of-view (FOV) is 15° by 15° which corresponds to 26.6 km by 26.6 km at 100 km altitude. The camera was oriented so that the FOV captured the magnetic zenith at the center. The relatively larger FOV as used by Kataoka et al. ([2011a], [2011b]) is one important factor for us to find simultaneous compound features.

In this letter, we show NIPR-CMOS data alone because these particular events did not show any features clearly faster than 25 Hz, which is the Nyquist frequency of NIPR-CMOS. A full-color DSLR camera (Nikon D4 with NIKKOR fish-eye 8 mm F2.8 lens, Nikon, Tokyo, Japan) was also installed alongside the CMOS cameras to provide an all-sky image every 10 s.

Two interesting examples are summarized in real-time playing (1 and 2). The quality of the obtained images is high enough to make the differential movies (3and 4), which are created by subtracting the previous image at each frame to emphasize the rapidly varying faint features. The differential movie technique is used to investigate the rapid variations in the auroral morphology, and it is another important factor for investigating auroral features as it draws attention to the flickering aurora events.

RESULTS

Event 1 (19 February 2014 09:30 UT) contains flickering rays and pulsating modulation at the middle of a surge in the pre-midnight sector, during the expansion phase of an intense substorm with the AE index >1,200 nT just after the storm peak of the Dst index = −112 nT. Event 2 (21 February 2014 12:01 UT) contains growing eddies at the poleward boundary of multiple arcs with localized flickering in the midnight sector, during the expansion phase of a moderate substorm with the AE index <500 nT in the storm recovery phase.

Event 1: Flickering Rays with Pulsating Modulation

Event 1 captures both flickering aurora and pulsating aurora in the pre-midnight sector at approximately 22 MLT. From Figure 1a, b, it is found that diffuse aurora appeared to the west, and discrete aurora appeared to the northeast at the middle of a surge. North-south aligned narrow black aurora is located between these two very different types of aurora. As seen in 1, and particularly in 3, the discrete aurora contains flickering rays, while the diffuse aurora consists of pulsating modulation. Figure 1c,d shows the time evolution along the white line in Figure 1a to partly represent the flickering (Figure 1d) and pulsating modulations (Figure 1c).

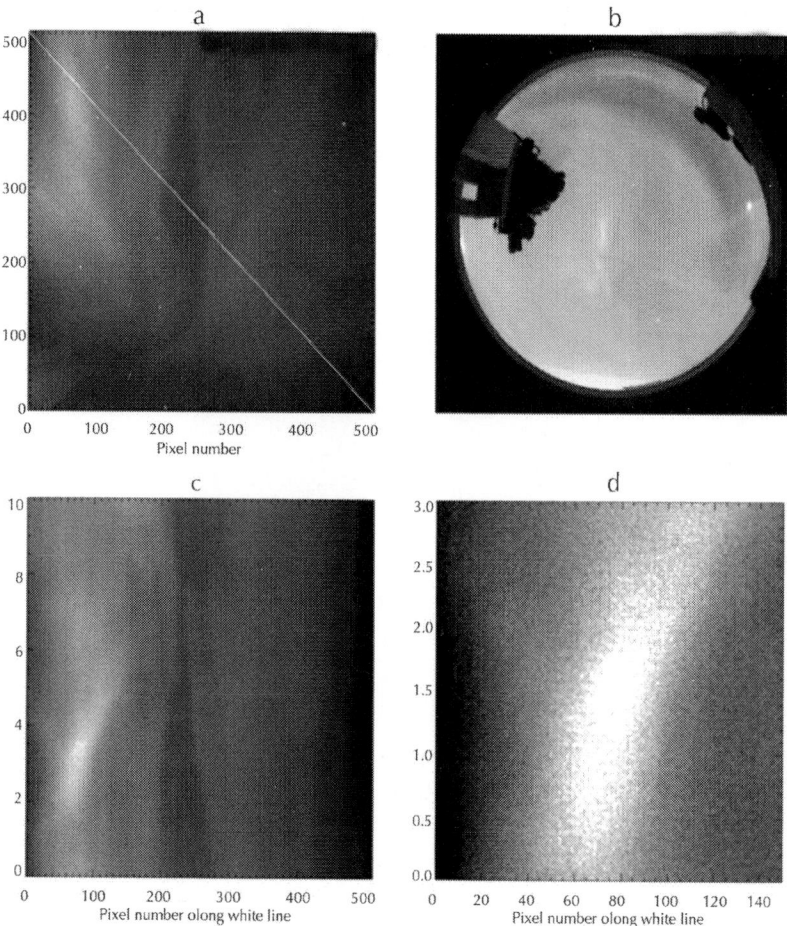

Figure 1: Pulsating aurora appeared to the west, and flickering rays appeared to the northeast. (a) Pulsating aurora appeared to the west, and flickering rays appeared to the northeast in the snapshot at 09:30:02 UT. North is to the top and west is to the right for sub-figures (a) and (b), and the magnetic zenith is at about the center of the image (a). (b)Full-color DSLR all-sky image at 09:30:00 UT shows that the compound structure is located at the middle of a surge. (c) Time evolution along the white line in the panel (a) during a 10-s time period from 09:30:00 UT and (d) the expanded view of the flickering rays is also shown.

Figure 2 shows the distribution map of Fourier spectral amplitude from 1 Hz to 24 Hz during the 1-s time interval from 09:30:03 UT to 09:30:04 UT on 19 February 2014, applying the same fast Fourier transform (FFT) analysis as described by Kataoka et al. ([2012]). The time series averaged over each 4 by 4 pixels is Fourier transformed, and the spectral amplitude distribution of the Fourier transform at each frequency is linearly gray scaled from black to white to see the spatial coverage. The flickering aurora appeared at 12 to 13 Hz in the rayed structure at the northeast area. Pulsating modulation is also found at 2 to 3 Hz as seen at the north area.

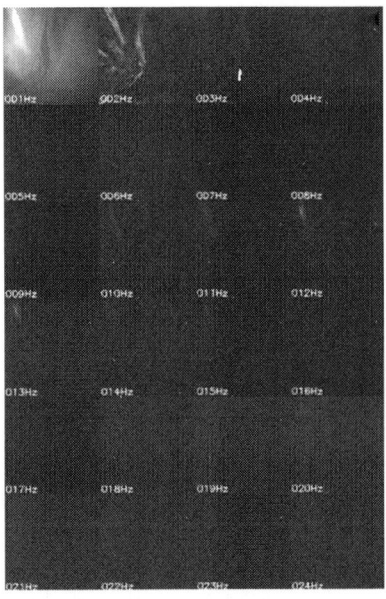

Figure 2: Fourier spectral intensity map at 09:30:03 to 09:30:04 UT on 19 February 2014. Flickering is clear at 12 to 13 Hz, while pulsating modulation is seen at 2 to 3 Hz.

Event 2: Growing Eddies with Localized Flickering

Figure 3 shows an example of growing eddies near the magnetic midnight sector at approximately 0.5 MLT. The east-west aligned arc is the poleward boundary of a multiple arc system as shown in Figure 3b. Wavy structure with a wavelength of approximately 5 km at 100 km altitude appeared at the poleward edge of the arc with 2 to 3 km width and gradually evolved into eddies of counterclockwise rotation in approximately 10 s. Small-scale vortices flow eastward at 12 to 20 km/s as seen in Figure 3d and the peaks of the poleward waves drift slowly eastward at an order of 0.1 km/s as shown in Figure 3c. Comparing Figure 3c, d during the 4-s time period from 12:01:11 UT to 12:01:15 UT, it is found that the eastward motion of small vortices along the white solid line is approximately 38 km/s and which is faster than the approximately 12 km/s eastward motion along the dashed line, i.e., eastward motion is faster at the poleward part of the arc. As shown in 2, the intensity of eddies is quasi-periodically enhanced at every 1 to 2 s by the passage of a series of bright small vortices flowing eastward in the arc.

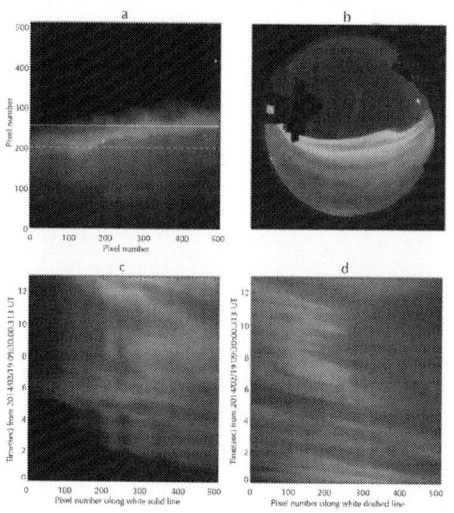

Figure 3: A snapshot of growing eddies is shown at 12:01:15 UT on 21 February 2014 (a). (b) Full-color DSLR all-sky image at 12:01:10 UT shows that the little wavy structure is located at the poleward boundary of an arc system. The formats are the same as Figure 1a, b for sub-figures (a) and (b). Time evolutions of the brightness along the white solid line and dotted line in panel (a) during a 13-s time period from 12:01:07 UT are shown in the panels (c) and (d), respectively.

By making the differential movie (4), it is found that flickering aurora transiently appeared during the growth of eddies at the poleward edge. The position of the relatively strong part of periodic oscillation is highlighted for every frequency in Figure 4, again applying the same FFT analysis as described by Kataoka et al. ([2012]). It is found that the flickering aurora appeared at below 7 Hz in a limited area along the arc. However, note that some of the apparent flickering in the FFT analysis could be due to spatial motion in the arc due to the combination of flickering and moving auroral features. This does not affect the conclusion that the flickering has frequencies up to 7 Hz. The strongest real flickering is on the left edge of the image at about 12:01:15 UT, and then in the center at about 12:01:17 UT.

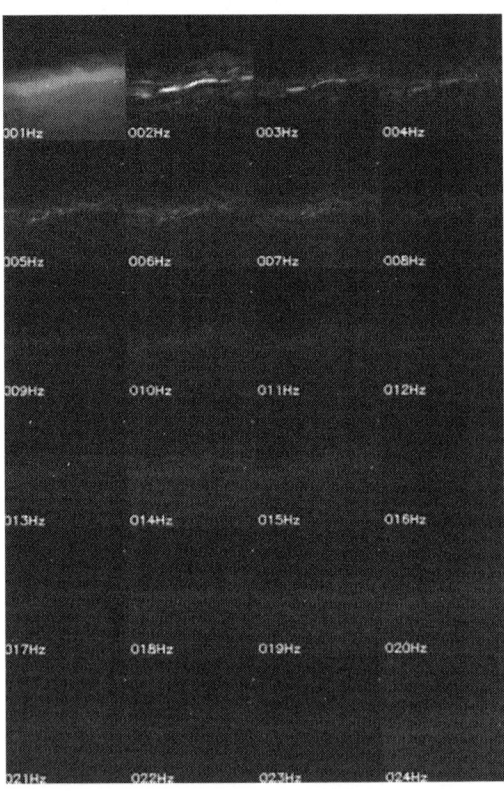

Figure 4: Fourier spectral intensity map at 12:01:15 to 12:01:16 UT on 21 February 2014. The fast variation of less than 7 Hz can be seen in the localized area within the arc, and there is no feature beyond 8 Hz.

DISCUSSION

In this report, we have briefly reported on observations of compound auroral microstructures using high-resolution optical instrumentation. In event 1, flickering rays appeared at the middle of surge beside a region of typical pulsating modulation, while in event 2, localized flickering aurora appeared associated with the growing eddies at the poleward edge of an arc system. We discuss the source mechanisms and the possible meanings of these two different compound microstructures.

For event 1, there is little doubt that the modulation region of the flickering aurora is in the magnetosphere-ionosphere coupled region as described in the 'Introduction' section, but the closely associated region of pulsating patches was not expected. Many recent studies have confirmed that post-midnight pulsating aurora and their fast modulations during the recovery phase of substorms originate from pitch-angle scattered energetic electrons from the magnetosphere by chorus waves (e.g., Miyoshi et al. [2010]; Nishiyama et al. [2012], [2014]). Different possible mechanisms of pulsating aurora also exist. For example, Sato et al. ([2004]) suggested that the modulation of the field-aligned potential drop may cause pulsating aurora. Because the observed pulsating aurora in this study appeared during the expansion phase of a substorm in the pre-midnight sector, the generation mechanism may be different from the typical post-midnight pulsating aurora during the recovery phase of substorms. It is therefore possible in this case that the source mechanisms of both flickering and pulsating modulation are common or closely involved to achieve the efficient way to dissipate the energy in the M-I coupled region. Further high-speed observations will be important to clarify whether such a coexistence of the flickering aurora and pulsating aurora is a common feature during the expansion phase of substorms.

Event 1 showed the flickering rays, which is consistent with the IAW-accelerated relatively low-energy electrons as suggested by Ivchenko et al. ([2005]). On the other hand, pulsating aurora, and especially the fast modulation, is the result of more energetic electrons (Sandahl et al. [1980]). The simplest interpretation of the coexistence would therefore be that both IAW and also energetic particles of pulsating modulation are enhanced during the expansion phase of the substorm to be launched all together into the ionosphere from the nearby locations in the magnetosphere.

Event 2 is consistent with the local source scenario of IAW (Asamura et al. [2009]), associated with the inverted-V type settings as inferred from the evolution of curls and folds (Chaston and Seki [2010]). Asamura et al. ([2009]) showed that vortical discrete auroral forms of approximately 8 km counter-stream at 14 to 18 km/s during the excitation of inertial Alfven waves at approximately 3,000 km altitude associated with the observed inverted-V electrons. It is therefore possible that the electrostatic shear-induced instability forming the curls and folds (Wagner et al. [1983]) is capable of launching IAWs (Asamura et al. [2009]) and thereby cause the observed localized flickering aurora. Relatively faint appearance of the localized flickering may be different from typical EMIC-related single-frequency flickering.

It is noteworthy that the growing eddies show counterclockwise rotation (tilted to the east), which is in the wrong direction as expected from the standard Kelvin-Helmholtz instability if we assume negligibly slow flow at poleward side with the fast eastward bulk flow in the arc. The flow shear within the arc is counterclockwise, i.e., the poleward part moves eastward faster than the equatorial part as described in the 'Event 2: growing eddies with localized flickering' section. The counterclockwise shear setting is consistent with the upward field-aligned current system associated with the locally converging electric field toward the arc. Nonlinear numerical simulations to reproduce the observed features of growing eddies are awaited in order to offer meaningful interpretation.

There are other potentially IAW-related auroras which should be discussed in some detail. Semeter and Blixt ([2006]) and Semeter et al. ([2008]) showed that 'arc packets' can be interpreted by the parallel electric field of IAW within the resonance cone. The arc packets occur especially at intense aurora with intense geomagnetic activity. It is noteworthy that Alfven wave propagation on sharp density gradients in a direction transverse to the static magnetic field leads to the formation of a significant parallel electric field (Genot et al. [2004]), which gives a spatiotemporal pattern similar to the arc packets. The relationship between arc packets and flickering has not been elucidated yet. It has been measured that field-aligned electron bursts (FABs) are superposed on mono-energetic inverted-V electrons (e.g., McFadden et al. [1987]), especially at the edge of the inverted-V, and the whole characteristics is similar to those of flickering aurora. Dahlgren et al. ([2013]) recently showed that 'flaming' aurora during auroral breakup is consistent with the FABs and is also consistent with 2.4 Hz flickering. In event 1, it is still difficult to tell the phase difference of flickering rays with altitude, which might be like flaming or time-of-flight dispersion. It is therefore important to further investigate the occurrence distribution of flaming, flickering, arc packets, and their compound features to distinguish the exact mechanisms and to identify their meaning via high-speed imaging observations. Continuous ground-based high-speed imaging observation of aurora at >100 fps is possible now (Kataoka et al. [2011a], [2011b]; Yaegashi et al. [2011]; Nishiyama et al. [2012], [2014]). It would be important to pursue the meanings via the occurrence distribution of the fine-scale rapidly varying aurora against meso-scale and global-scale auroral dynamics.

The next important challenge of high-speed optical observations would be the systematic survey of the fastest variation appearing in aurora including the helium and hydrogen cyclotron frequency. Yaegashi et al. ([2011]) suggested that existence of flickering aurora is related to helium ion EMIC waves. Temerin et al. ([1986]) originally suggested that what appears to be a steady field-aligned flux of auroral electrons can also be produced by hydrogen cyclotron waves since modulation at the hydrogen cyclotron frequency is hardly resolved by current typical sampling rate of 40 ms (25 Hz) of particle detectors onboard Reimei satellite, for example. In fact, McFadden et al. ([1987]) reported the *in situ* observational results of wave-particle interactions between H+ EMIC waves (at approximately 120 Hz) and down-going field-aligned electron fluxes in inverted V arcs at approximately 3,700 km altitude. McHarg et al. ([1998]) reported up to 180 Hz variation of aurora using ground-based high-speed photometer observations.

AUTHORS' CONTRIBUTIONS

RK designed the camera system, carried out the observation, and drafted the manuscript. YF performed the data analysis. YM helped to design the camera system and to draft the manuscript. HM, SI, and DH helped to install the camera system. YE developed the program to control the camera system. HD, DW, and NI advised the interpretation and helped to draft the manuscript. All authors read and approved the final manuscript.

ACKNOWLEDGEMENTS

R. K. thanks Kevin Abnett, Naoki Sunagawa, and Ayumi Hashimoto for their help to installing the sCMOS camera systems. RK thanks Tomonao Harada and Nikon professional services for their support by providing Nikon cameras for our observations. This work is supported by Yamada Science Foundation and Grants-in-Aid for Scientific Research (19403010; 25302006) from the Ministry of Education, Culture, Sports, Science and Technology of Japan.

REFERENCES

1. Asamura K, Chaston CC, Itoh Y, Fujimoto M, Sakanoi T, Ebihara Y, Yamazaki A, Hirahara M, Seki K, Kasaba Y, Okada M (2009) Sheared flows and small-scale Alfven wave generation in the auroral acceleration region. Geophys Res Lett 36:L05105 doi:10.1029/2008GL036803

2. Chaston CC, Seki K (2010) Small-scale auroral current sheet structuring. J Geophys Res 115:A11221 doi:10.1029/2010JA015536

3. Chen L-J, Kletzing CA, Hu S, Bounds SR (2005) Auroral electron dispersion below inverted-V energies: resonant deceleration and acceleration by Alfven waves. J Geophys Res 110:A10S13 doi:10.1029/2005JA011168

4. Dahlgren H, Ivchenko N, Lanchester BS, Sullivan J, Marklund G, Whiter D (2008) Morphology and dynamics of aurora at fine scale: first results from the ASK instrument. Ann Geophys 26:1041-1048 doi:10.5194/angeo-26-1041-2008

5. Dahlgren H, Ivchenko N, Lanchester BS, Sullivan BSJ, Whiter D, Marklund G, Stromme A (2008) Using spectral characteristics to interpret auroral imaging in the 731.9 nm O+ line. Ann Geophys 26:1905-1917 doi:10.5194/angeo-26-1905-2008

6. Dahlgren H, Aikio A, Kaila K, Ivchenko N, Lanchester BS, Whiter DK, Marklund GT (2010) Simultaneous observations of small multi-scale structures in an auroral arc. J Atmos Sol Terr Phys 72:633 doi:10.1016/j.jastp.2010.01.014

7. Dahlgren H, Ivchenko N, Lanchester BS (2012) Monoenergetic high-energy electron precipitation in thin auroral filaments. Geophys Res Lett 39:L20101 doi:10.1029/2012GL053466

8. Dahlgren H, Semeter JL, Marshall RA, Zettergren M (2013) the optical manifestation of dispersive field-aligned bursts in auroral breakup arcs. J Geophys Res Space Physics 118:4572-4582 doi:10.1002/jgra.50415

9. Genot V, Louarn P, Mottez F (2004) Alfven wave interaction with inhomogeneous plasmas: acceleration and energy cascade towards small-scales. Ann Geophys 22:2081-2096

10. Goertz CK, Boswell RW (1979) Magnetosphere-ionosphere coupling. J Geophys Res 84(A12):7239-7246 doi:10.1029/JA084iA12p07239

11. Gustavsson B, Lunde J, Blixt EM (2008) Optical observations of flickering aurora and its spatiotemporal characteristics. J Geophys Res 113:A12317 doi:10.1029/2008JA013515

12. Hallinan TJ (1976) Auroral Spirals 2. Theory. J Geophys Res 81(22):3959-3965

13. Hallinan TJ, Davis TN (1970) Small-scale auroral arc distortions. Planet Space Sci 18:1735-1744

14. Hasegawa A (1976) Particle acceleration by MHD surface wave and formation of aurora. J Geophys Res 81(28):5083-5090 doi:10.1029/JA081i028p05083

15. Ivchenko N, Blixt EM, Lanchester BS (2005) Multispectral observations of auroral rays and curls. Geophys Res Lett 32:L18106 doi:10.1029/2005GL022650

16. Kataoka R, Miyoshi Y, Sakanoi T, Yaegashi A, Shiokawa K, Ebihara Y (2011a) Turbulent microstructures and formation of folds in auroral breakup arc. J Geophys Res 116:A00K02, doi:10.1029/2010JA016334

17. Kataoka R, Miyoshi Y, Sakanoi T, Yaegashi A, Ebihara Y, Shiokawa K (2011) Ground-based multispectral high-speed imaging of flickering aurora. Geophys Res Lett 38:L14106 doi:10.1029/2011GL048317

18. Kataoka R, Miyoshi Y, Hampton D, Ishii T, Kozako H (2012) Pulsating aurora beyond the ultra-low-frequency range. J Geophys Res 117:A08336 doi:10.1029/2012JA017987

19. Kunitake M, Oguti T (1984) Spatial-temporal characteristics of flickering spots in flickering auroras. J Geomagn Geoelectr 36:121

20. Lanchester BS, Rees MH, Lummerzheim D, Otto A, Frey HU, Kaila KU (1997) Large fluxes of auroral electrons in filaments of 100 m width. J Geophys Res 102(A5):9741-9748 doi:10.1029/97JA00231

21. Lanchester BS, Ashrafi M, Ivchenko N (2009) Simultaneous imaging of aurora on small scale in OI (777.4 nm) and N21P to estimate energy and flux of precipitation. Ann Geophys 27:2881-2891 doi:10.5194/angeo-27-2881-2009

22. McFadden JP, Carlson CW, Boehm MH, Hallinan TJ (1987) Field-aligned electron flux oscillations that produce flickering aurora. J Geophys Res 92(A10):11133-11148 doi:10.1029/JA092iA10p11133

23. McHarg MG, Hampton DL, Stenbaek-Nielsen HC (1998) Fast photometry of flickering in discrete auroral arcs. Geophys Res Lett 25(14):2637-2640

24. Michell RG, McHarg MG, Samara M, Hampton DL (2012) Spectral analysis of flickering aurora. J Geophys Res 117:A03321 doi:10.1029/2011JA016703

25. Miyoshi Y, Katoh Y, Nishiyama T, Sakanoi T, Asamura K, Hirahara M (2010) Time of flight analysis of pulsating aurora electrons, considering wave-particle interactions with propagating whistler mode waves. J Geophys Res 115:A10312 doi:10.1029/2009JA015127

26. Nishiyama T, Sakanoi T, Miyoshi Y, Kataoka R, Hampton D, Katoh Y, Asamura K, Okano S (2012) Fine scale structures of pulsating auroras in the early recovery phase of substorm using ground-based EMCCD camera. J Geophys Res 117:A10229 doi:10.1029/2012JA017921

27. Nishiyama T, Sakanoi T, Miyoshi Y, Hampton D, Katoh Y, Kataoka R, Okano S (2014), Multi-scale temporal variations of pulsating auroras: on-off pulsation and a few-Hz modulation, J Geophys Res, 119, doi:10.1002/2014JA019818.

28. Royrvik O, Davis TN (1977) Pulsating aurora: local and global morphology. J Geophys Res 82(29):4720-4740

29. Sakanoi K, Fukunishi H, Kasahara Y (2005) A possible generation mechanism of temporal and spatial structures of flickering aurora. J Geophys Res 110:A03206 doi:10.1029/2004JA010549

30. Sandahl I, Eliasson L, Lundin R (1980) Rocket observations of precipitating electrons over a pulsating aurora. Geophys Res Lett 7:309-312 doi:10.1029/GL007i005p00309

31. Sandahl I, Sergienko T, Brandstrom U (2008) Fine structure of optical aurora. J Atmos Sol Terr Phys 70:2275-2292 doi:10.1016/j.jastp.2008.08.016

32. Sato N, Wright DM, Carlson CW, Ebihara Y, Sato M, Saemundsson T, Milan SE, Lester M (2004) Generation region of pulsating aurora obtained simultaneously by the FAST satellite and a Syowa-Iceland conjugate pair of observatories. J Geophys Res 109:A10201 doi:10.1029/2004JA010419

33. Semeter J, Blixt EM (2006) Evidence for Alfven wave dispersion identified in high-resolution auroral imagery. Geophys Res Lett 33:L13106 doi:10.1029/2006GL026274

34. Semeter J, Zettergren M, Diaz M, Mende S (2008) Wave dispersion and the discrete aurora: new constraints derived from high-speed imagery. J Geophys Res 113:A12208 doi:10.1029/2008JA013122

35. Temerin M, McFadden J, Boehm M, Carlson CW, Lotko W (1986) Production of flickering aurora and field-aligned electron flux by electromagnetic ion cyclotron waves. J Geophys Res 91(A5):5769-5792 doi:10.1029/JA091iA05p05769

36. Temerin M, Carlson C, McFadden JP (1993) The acceleration of electrons by electromagnetic ion cyclotron waves, 'Auroral Plasma Dynamics', Geophys. Monogr. Ser., 80, Lysak RL (ed) AGU, Washington, D. C. pp. 155–161, doi:10.1029/GM080p0155

37. Wagner JS, Sydora RD, Tajima T, Hallinan T, Lee LC, Akasofu S-I (1983) Small-scale auroral arc deformations. J Geophys Res 88(A10):8013-8019 doi:10.1029/JA088iA10p08013

38. Whiter DK, Lanchester BS, Gustavsson B, Ivchenko N, Sullivan JM, Dahlgren H (2008) Small-scale structures in flickering aurora. Geophys Res Lett 35:L23103 doi:10.1029/2008GL036134

39. Whiter DK, Lanchester BS, Gustavsson B, Ivchenko N, Dahlgren H (2010) Using multispectral optical observations to identify the acceleration mechanism responsible for flickering aurora. J Geophys Res 115:A12315 doi:10.1029/2010JA015805

40.   Yaegashi A, Sakanoi T, Kataoka R, Asamura K, Miyoshi Y, Sato M, Okano S (2011) Spatial-temporal characteristics of flickering aurora as seen by high-speed EMCCD imaging observations. J Geophys Res 116:A00K04 doi:10.1029/2010JA016333

Study of Pc1 Pearl Structures Observed at Multi-Point Ground Stations in Russia, Japan, and Canada

Chae-Woo Jun[1], Kazuo Shiokawa[1], Martin Connors[2],
Ian Schofield[2], Igor Poddelsky[3], and Boris Shevtsov[3]

[1]Solar-Terrestrial Environment Laboratory, Nagoya University, Furo-cho, Chikusa-ku, Nagoya 464-8601, Japan

[2]Center for Science, Athabasca University, 1 University Drive, Athabasca AB T9S 3A3, Canada

[3]Institute of Cosmophysical Research and Radiowave Propagation, Far Eastern Branch of the Russian Academy of Sciences, 7 Mirnaya Street, Paratunka 684034, Kamchatka region, Russian Federation

ABSTRACT

We investigate possible generation mechanisms of Pc1 pearl structures using multi-point induction magnetometers in Athabasca in Canada,

Magadan in Russia, and Moshiri in Japan. We selected two Pc1 pulsations that were simultaneously observed at the three stations and applied a polarization analysis. In case 1, on 8 April 2010, Pc1 pearl structures were slightly different in some time intervals at different stations, and their polarization angles varied depending on the frequencies at the three stations. Case 2, on 11 April 2010, showed Pc1 pearl structures that were similar at different stations, and their polarization angle was independent of frequency at all three stations. In order to understand these differences, we performed two simple model calculations of Pc1 pearl structures under different conditions. The first model assumes that Pc1 waves propagated from a latitudinally extended source with different frequencies at different latitudes to the observation points, representing beating of these waves in the ionosphere. The second model considers Pc1 waves for which different frequencies are mixed at a point source to cause the beating at the source point, indicating that the Pc1 pearl structures are generated in the magnetosphere. The first model shows slightly different waveforms at different stations. In contrast, the second model shows identical waveforms at different stations. From these results, we conclude that, in case 1, Pc1 pearl structures were caused by beating in the ionosphere. On the other hand, in case 2, they were the result of magnetospheric effects. We suggest that beating processes in the ionosphere could be one of the generation mechanisms of Pc1 pearl structures.

BACKGROUND

Electromagnetic ion cyclotron (EMIC) waves are known to be generated in the equatorial region of the magnetosphere at L of approximately 4 to 8 (Anderson et al. [1992]) due to ion cyclotron instability of energetic resonant ions with temperature anisotropy. These waves propagate along the geomagnetic field lines to the ionosphere, where they interact with the ionospheric plasma, generating compressional-mode waves. EMIC waves can generate isolated proton auroras via wave-particle interactions at subauroral latitudes (e.g., Sakaguchi et al. ([2008]); Nomura et al. ([2012])). These ionospheric waves are observed as Pc1 geomagnetic pulsations in the frequency range of 0.2 to 5.0 Hz. The characteristics of Pc1 pulsations in dynamic spectra have been classified by Fukunishi et al. ([1981]). Pc1 pulsations can be trapped in the duct of the ionospheric F layer and propagate horizontally, from high to low latitudes, over long distances of several thousand kilometers (e.g., Manchester ([1966]); Tepley and Landshoff ([1966]); Campbell ([1967]); Kuwashima et al. ([1981]); Kawamura et al. ([1981]); Kim et al. ([2011]); Waters et al. ([2013])).

Pc1 pearl structures are amplitude modulations of Pc1 waves, which show a quasi-periodic intensification of amplitude with a repetition period of several tens of seconds (Troitskaya and Gul'Elmi [1967]). Pc1 pearl structures are the most common form of Pc1 pulsations. They have been studied for many years using ground-based and satellite observations, in order to understand the generation mechanisms of Pc1 pearl structures (e.g., Perraut ([1982]); Erlandson et al. ([1990]); Guglielmi et al. ([1996]); Mursula ([2007]); Rasinkangas and Mursula ([1998]); Mursula et al. ([1999]); Mursula ([2007]); Usanova et al. ([2008]); Nomura et al. ([2011])). However, the generation mechanisms of Pc1 pearl structures have yet to be clearly identified. For more than 50 years, the bouncing wave packets (BWP) model has been believed to be a possible generation mechanism of Pc1 pearl structures (e.g., Jacobs and Watanabe ([1964]); Obayashi ([1965])). This model suggests that Pc1 pearl structures are caused by the bouncing of Pc1 waves along the geomagnetic field lines, between the northern and southern hemispheres. Mursula et al. ([1999]) found that there was a strong negative correlation between the repetition period of Pc1 pearl structures and the frequency of Pc1 pulsations. They implied that the L-shell dependence of various Pc1 frequencies, and the repetition period of Pc1 pearl structures may be interpreted by the BWP model. Nonetheless, recent studies have begun to consider whether or not the BWP model is the best mechanism to explain Pc1 pearl structures. For example, the fact that the repetition period of Pc1 pearl structures observed on the ground is shorter than that expected from the BWP model (e.g., Nomura et al. ([2011]); Usanova et al. ([2008])) suggests that this model might not be entirely valid. Another possible generation mechanism of Pc1 pearl structures in the magnetosphere is the modulation of EMIC waves by long-period ultra-low frequency (ULF) waves, such as Pc4-5 pulsations (e.g., Mursula et al. ([1997]); Rasinkangas and Mursula ([1998]); Mursula et al. ([2001]); Mursula ([2007])). Mursula et al. ([1997]) found a long Pc1 pulsation event on 11 April 1986 observed by the Finnish magnetometer network and the Viking satellite. They suggested that the pearl structures are modulated Pc1 waves propagating within or at the plasmapause, and that hydromagnetic chorus events consist of similar wave packets propagating outside the plasmasphere. Moreover, Rasinkangas and Mursula ([1998]) found that the EMIC waves observed by the Viking satellite in the inner magnetosphere were modulated by magnetospheric Pc3 pulsations. Hence, they proposed that the modulation of EMIC waves by long-period ULF waves could be an alternative to the BWP model. Other studies have also begun to consider ionospheric effects as a possible generation mechanism for Pc1 pearl structures. Pope ([1964]) and Nomura et al. ([2011]) suggested that beating in the ionosphere could create Pc1 pearl structures. Pope ([1964]) suggested that Pc1 waves with different frequencies are combined during their propagation through the ionospheric duct. Nomura et al. ([2011]) found that approximately 70 %of the Pc1 pulsation events observed at low-latitude ground stations have a frequency dependence on the polarization angle. They inferred that these Pc1 events, coming from a spatially distributed ionospheric source, can create pearl structures by the beating of Pc1 waves with slightly different frequencies in the ionosphere. However, previous studies have not investigated what type of generation mechanisms, either in the magnetosphere or in the ionosphere, contribute more to the formation of Pc1 pearl structures.

In this study, we report two Pc1 pulsation events simultaneously observed at longitudinally and latitudinally separated ground stations. In order to consider Pc1 pulsations coming from the same source region, observed simultaneously at three stations, we used a coherence analysis of their waveforms. Moreover, we investigated the similarity of Pc1 pearl structures between two stations using a cross correlation analysis of the upper envelopes of Pc1 pearl structures. Details of the analyses are presented in the 'Observations' subsection. We found two Pc1 pulsations simultaneously observed at these three stations. In case 1, the Pc1 pearl structures are slightly different at different stations, whereas in case 2, they are similar at different stations. According to our model calculations of Pc1 pearl structures, we found that a spatially distributed ionospheric source can create different waveforms at different stations. This research is the first case study comparing the similarity of Pc1 pearl structures at very distant ground stations. This research provides evidence suggesting that beating in the ionosphere is one of the possible generation mechanisms of Pc1 pearl structures.

METHODS

Observations

We investigated Pc1 pearl structures using multi-point induction magnetometers deployed by the Solar-Terrestrial Environment Laboratory, Nagoya University, at Athabasca (ATH) in Canada, Magadan (MGD) in Russia, and Moshiri (MOS) in Japan. Figure 1 shows the locations of these ground stations, giving their geographic latitudes and longitudes, as well as dipole geomagnetic latitudes, calculated using the IGRF-11 model with an epoch time of 2010. The distance between ATH and MGD is about 4,000 km, while MGD and MOS are separated by about 1,000 km. In this study, we use the data giving the H, D, and Z geomagnetic field components with a sampling rate of 64 Hz and a GPS clock accuracy of about 1 μs. The magnetometer sensors have almost identical sensitivities and provide phase differences for H, D, and Z components in the Pc1 frequency range (0.2 to 5.0 Hz). The magnetometer data, originally in volts, were converted to nT units by considering the frequency-dependent amplitude of the system. The sensor sensitivity was measured by a calibration coil with a length of 2 m. Additional details of these induction magnetometers are given by Shiokawa et al. ([2010]).

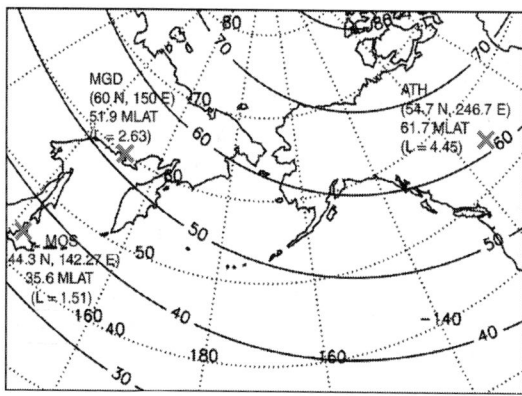

Figure 1: Location of the three induction magnetometer stations. These stations are located at Athabasca (ATH) in Canada, Magadan (MGD) in Russia,

and Moshiri (MOS) in Japan. Solid lines indicate dipole magnetic latitudes calculated using the IGRF-11 model with an epoch time of 2010. Dashed lines indicate geographic coordinates.

Data Analysis and Event Selection

The power spectrum density (PSD) of the two horizontal magnetic field components, H and D, measured by the induction magnetometers, was calculated by fast Fourier transform (FFT) every 30 s. We used a time window of 128 s, with a frequency resolution of 0.0078 Hz. In each window, the polarization angle is calculated using the relationship described in Fowler et al. ([1967]). The polarization angle is defined as the positive (negative) angle measured from the magnetic north westward (eastward). The coherence of the Pc1 waveforms helps us to determine whether or not the Pc1 pulsations come from the same source region in the frequency domain. If the coherence of Pc1 waveforms is close to one, with the same frequencies between two signals, the two signals are identical, indicating that these two signals are coming from the same source. On the other hand, if the coherence is equal to zero, the two signals have no correlation, indicating that these two signals are coming from two different sources.

From 1 January 2009 to 31 December 2011, we investigated three years of dynamic spectral data of geomagnetic pulsations obtained at ATH, MGD, and MOS. We considered four criteria in selecting Pc1 pulsations by visual inspection of the spectra: Pc1 pulsations are well-defined in the PSD and within the frequency range of 0.2 to 5.0 Hz; the Pc1 pulsation power should be greater than 10^{-4}nT ^{2}/Hz; the Pc1 frequency bandwidth should be wider than 0.2 Hz; the Pc1 pulsation should be discernible for more than 20 min.

Using these criteria, we selected 509 Pc1 events at ATH, 366 at MGD, and 518 at MOS. Then, we chose 28 Pc1 pulsations that were observed simultaneously at these three stations. We defined those events that have a coherence value greater than 0.5 as high coherence events. The number of high coherence events observed between ATH and MGD was seven. Only two of those events were observed clearly at all three stations without contamination by other frequency bands of Pc1 pulsations. We selected two Pc1 pulsations that show similar shapes in the dynamic spectra at all three stations: case 1 on 8 April

2010, and case 2 on 11 April 2010.

RESULTS

Case 1 - 8 April 2010

Figure 2 shows the dynamic spectra of H and D components of the magnetic field variations, polarization angle, coherence between different stations, and cross-correlation of Pc1 amplitude envelopes (red lines in Figure 3), for a clear Pc1 geomagnetic pulsation event observed simultaneously at ATH, MGD, and MOS at 10:00 to 12:00 UT on 8 April 2010. We chose the cross-correlation between the ATH H component and the MGD D component, since they show the highest correlation between ATH and MGD. During this event, geomagnetic activity was slightly elevated (K_p = 1 to 2), with AE indices of approximately 300 to 500 nT. The local time at ATH was 00:00 to 02:00 LT, and the local time at MGD and MOS was 06:00 to 08:00 LT. In Figure 2a,b,c,d,e,f, Pc1 pulsations can be clearly identified at the three stations in the frequency range of 0.4 to 1.2 Hz. The intensity of PSD at MOS is weaker than at the other two stations, probably because of attenuation due to ionospheric duct propagation to lower latitudes (Kawamura et al. [1981]). We can see that the Pc1 pulsations are observed during three separated time intervals: 10:12 to 10:33, 10:41 to 10:53, and 10:55 to 11:49 UT at all stations. At ATH, the first and second Pc1 bursts can be clearly seen in Figure 2a,b, but the third Pc1 burst is much weaker than the others. On the other hand, Pc1 pulsations are clearly identifiable at MGD and MOS for all three intervals.

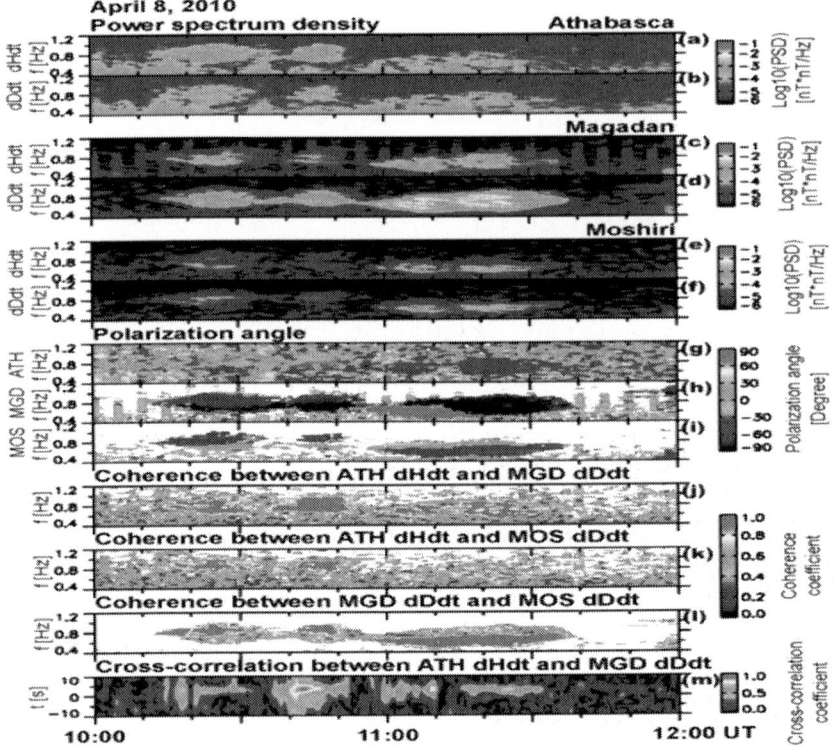

Figure 2: Spectral information in case 1. Power spectrum density of the (a) H and (b) D components of magnetic field at ATH, (c) H and (d)D components of magnetic field at MGD, and (e) H and (f) D components of magnetic field at MOS; polarization angle at (g) ATH, (h) MGD, and (i)MOS; coherence between (j) ATH H and MGD D, (k) ATH H and MOS D, and (l) MGD D and MOS D components; (m) cross correlation of the upper envelopes of Pc1 pearl structures between ATH H and MGD D components observed at 10:00 to 12:00 UT on 8 April 2010, for a frequency range of 0.4 to 1.2 Hz.

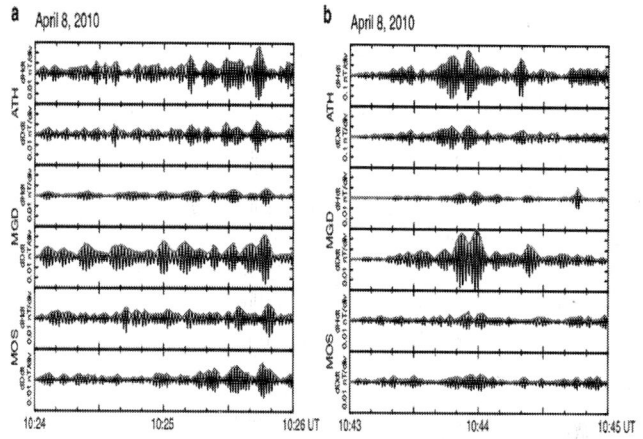

Figure 3: Time series analysis in case 1. (Top to bottom) band-pass filtered (0.5 to 1.2 Hz) Pc1 waveform of the magnetic field of the H and D components observed at ATH, MGD, and MOS at (a) 10:24 to 10:26 and(b) 10:43 to 10:45 UT on 8 April 2010. Red solid lines indicate upper envelope of Pc1 pearl structures.

The polarization angle at ATH, MGD, and MOS is shown in Figure 2g,h,i. For ATH (Figure 2g), it increased from approximately −50° (dark blue) to −20° (light blue) for 0.6 to 1 Hz. At MGD (Figure2h), the polarization angle varies from approximately −90° (black) to −50° (dark blue) in the same frequency range as ATH. The frequency dependence at MOS (Figure 2i) shows a decrease of the polarization angle from approximately 30° (green) to −50° (dark blue). According to Nomura et al. ([2011]), this polarization angle dependence on frequency indicates that the Pc1 pulsation observed at the three stations at 10:00 to 12:00 UT has a spatially distributed ionospheric source at high latitudes.

In order to distinguish whether the Pc1 pulsations propagated from the same ionospheric source, we show the coherence of Pc1 waveforms between each pair of stations in Figure 2j,k,l. High coherence of Pc1 waveforms was observed between two stations, indicating that the Pc1 pulsations observed at the three different stations were propagated from the same origin in the ionosphere.

Figure 3a,b shows the waveforms of the H and D components of Pc1 pulsations at ATH, MGD, and MOS on 8 April 2010 at 10:24 to 10:26 and 10:43 to 10:45 UT, respectively. In order to remove noise from other frequencies, we show the time series of Pc1 pulsations obtained using a band-pass filter in the frequency range of 0.5 to 1.2 Hz. The amplitude modulation of the pulsations varies with a repetition period of approximately 10 s in both time intervals. The time difference between ATH and MGD (MGD and MOS) is approximately 4 s (0 s) and was confirmed by lag correlation studies (see below). The repetition periods of pearl structure changes in time during this event at all stations, and the ones observed at ATH, MGD, and MOS are generally similar but differ in their details.

Figure 4 shows the PSD of the H and D components of magnetic field variations at ATH and MGD, respectively, on 8 April 2010 and the coherence of Pc1 waveforms between these two components at 10:24:00 to 10:26:08 (Figure 4a,b,c) and 10:43:00 to 10:45:08 UT (Figure 4c,d,e) with a resolution of 0.0078 Hz. The time intervals of Figure 4a,b,c,d,e correspond to those shown in Figure 3a,b. In Figure 4a,b and Figure 4d,e, we can see a continuous high-PSD band at frequencies of 0.6 to 1.0 and 0.7 to 1.0 Hz, respectively. The coherence between the H component at ATH and the D component at MGD is close to one in the latter frequency range.

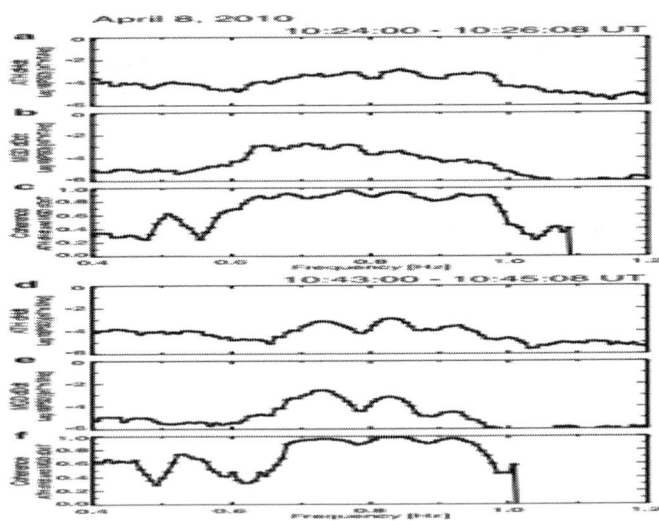

Figure 4: The PSD and the coherence of Pc1 waveforms of the H and D components in case 1. Upper three panels show the power spectrum density of (a) H component of the magnetic field at ATH and (b) D component of the

magnetic field at MGD, as well as (c) coherence of Pc1 waveforms between the H component at ATH and D component at MGD, observed at 10:24:00 to 10:26:08 UT on 8 April 2010, for a frequency range of 0.4 to 1.2 Hz. Lower three panels (d-f) show the same quantities observed at 10:43:00 to 10:45:08 UT on 8 April 2010.

We investigated the cross-correlation between the H component at ATH and the D component at MGD using the upper envelope of Pc1 pearl structures in the time domain. We used only these components because they have the highest PSD intensity compared with the other components and the background intensity. Figure 5a,b shows the cross-correlation between the H component at ATH and D component at MGD obtained for the upper envelope of the Pc1 pearl structure at 10:24:00 to 10:26:08 and 10:43:00 to 10:45:08 UT, respectively, on 8 April 2010. In Figure 5a, we can see that the correlation is greater than 0.5 with a time difference of approximately 3 s. On the other hand, in Figure 5b, the cross-correlation between ATH and MGD is close to 0.9, indicating that the upper envelope of the Pc1 pearl structures is generally similar, with a time difference of approximately 3 s. In Figure 2m, the cross-correlation between the H component at ATH and the D component at MGD is greater than 0.5 throughout this event. Especially, for the second of the Pc1 bursts, the coherence and correlation are both extremely high ($r > 0.8$). For the first and third Pc1 burst time intervals, however, the correlation decreases to 0.5, even if the coherence of the Pc1 waveforms is clearly close to one.

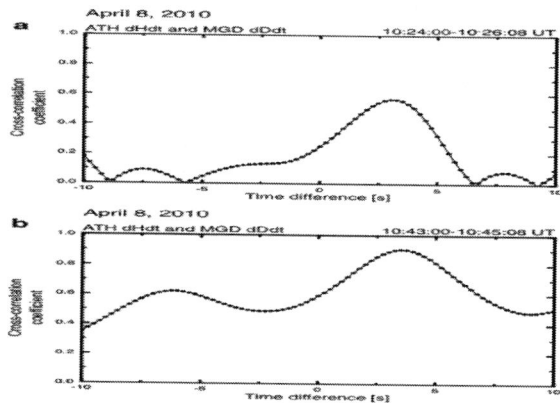

Figure 5: The cross-correlation analysis in case 1. Cross-correlation between

H component at ATH and D component at MGD, obtained using upper envelope of Pc1 pearl structures at (a) 10:24:00 to 10:26:08 and(b) 10:43:00 to 10:45:08 UT on 8 April 2010.

Case 2 - 11 April 2010

Figure 6 shows the dynamic spectra of the H and D components of magnetic field variations, polarization angle, coherence between different stations, and cross-correlation of Pc1 amplitude envelopes (red lines in Figure 7) between the ATH H component and the MGD D component, for a clear Pc1 geomagnetic pulsation event observed simultaneously at ATH, MGD, and MOS at 11:00 to 13:00 UT on 11 April 2010. Geomagnetic activity was relatively high during this event, with K_p = 3 to 4. The average AE index during the 11:00 to 13:00 UT interval was approximately 104 nT. The local time at ATH was 01:00 to 03:00 LT, and 07:00 to 09:00 LT at MGD and MOS. In Figure6a,b,c,d, Pc1 pulsations were clearly identified at ATH and MGD in the frequency range of 0.2 to 0.8 Hz. However, the Pc1 pulsations observed at MOS were much weaker than those seen at the other two stations (Figure 6e,f). At 11:30 to 12:20 UT at ATH, we can see three different frequency bands of Pc1 pulsations at 0.38 to 0.48, 0.5 to 0.6, and 0.61 to 0.73 Hz. At MGD, the two different frequency bands of Pc1 pulsations at 0.4 to 0.47 and 0.52 to 0.6 Hz can be seen in the D component of magnetic field variation (Figure 6d).

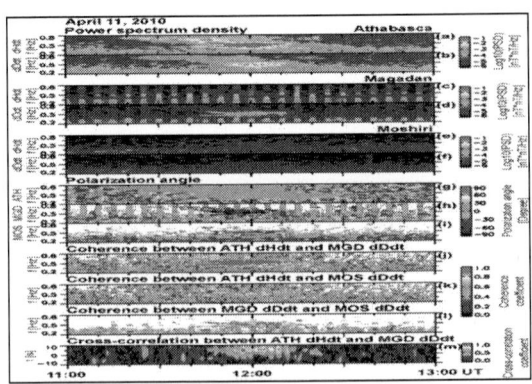

Figure 6: Spectral information in case 2. Power spectrum density of(a) H and (b) D components of magnetic field at ATH, (c)H and (d) D components of

magnetic field at MGD, and (e) H and (f) D components of magnetic field at MOS; (g) polarization angle at ATH, (h) MGD, and (i)MOS; coherence between (j) ATH H and MGD D, (k) ATH H and MOS D, and (l)MGD D and MOS D components; (m) cross correlation of the upper envelopes of Pc1 pearl structures between ATH H and MGD D components observed at 11:00 to 13:00 UT on 11 April 2010, for a frequency range of 0.2 to 0.8 Hz.Vertical stripes seen every approximately 5 s in the PSD at MGD are due to artificial noise.

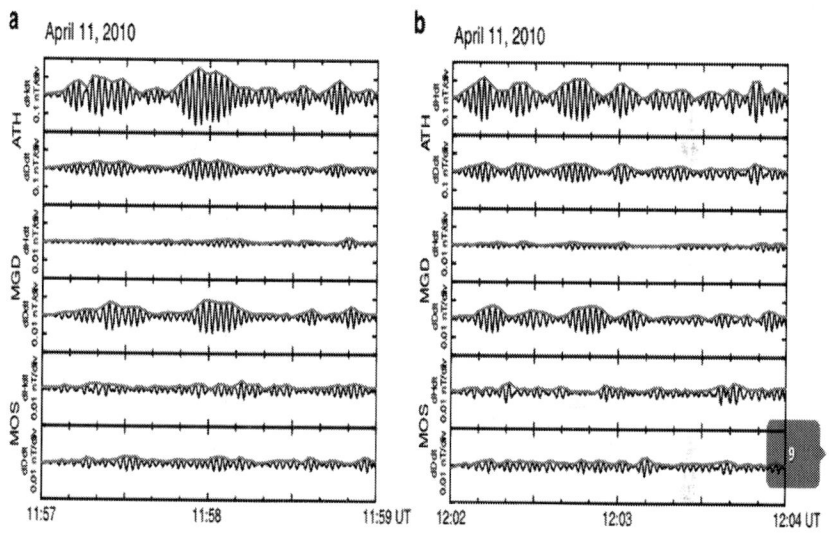

Figure 7: Time series analysis in case 2. (Top to bottom) band-pass filtered (0.3 to 0.7 Hz) Pc1 waveforms of the magnetic field of H and D components observed at ATH, MGD, and MOS at (a) 11:57 to 11:59 and(b) 12:02 to 12:04 UT on 11 April 2010. Red solid lines indicate upper envelope of Pc1 pearl structures.

The polarization angles at ATH, MGD, and MOS are shown in Figures 6g-6i. From 0.3 Hz to 0.7 Hz, the polarization angle at ATH (Figure 6g) is almost constant, with values around 0° (light green). In Figure 6h, the polarization angle at MGD barely varies, and remains near −60° (dark blue) in the frequency range of 0.3 to 0.6 Hz. The frequency dependence at MOS, as seen in Figure 6i, is not clear because the intensity of the Pc1 pulsations is very weak. The constant polarization angle observed at ATH and MGD, independent of frequency, suggests that Pc1 pulsations have a localized ionospheric source at high latitudes (Nomura et al. [2011]).

In order to distinguish whether Pc1 pulsations propagated from the same ionospheric source, we show the coherence of Pc1 waveforms between the different stations in Figure 6j,k,l. The coherence between two stations is close to one for all frequencies, even if the Pc1 pulsation at MOS was weak for frequencies near 0.5 Hz. This indicates that the Pc1 pulsations observed at the three different stations propagated from the same origin.

Figure 7a,b shows the waveforms of the H and D components of Pc1 pulsations observed at ATH, MGD, and MOS on 11 April 2010 at 11:57 to 11:59 and 12:02 to 12:04 UT, respectively. We show the time series of Pc1 pulsations obtained using a band-pass filter from 0.3 to 0.7 Hz in order to remove noise from other frequencies. Even if the amplitude modulation of Pc1 pulsations at MOS was weak in both time intervals, we can clearly see Pc1 pearl structures in all three stations. The repetition periods of Pc1 pearl structures vary from 10 to 40 s. We also note that there is a time difference of a few seconds in the pearl structures between ATH and MGD. The Pc1 pearl structures observed at ATH and MGD are similar in the time series of magnetic field variation, even though these two stations are separated by a distance of approximately 4,000 km.

Figure 8 shows the PSD of the H and D component magnetic field variations at ATH and MGD, respectively, and the coherence between these two components at 11:57:00 to 11:59:08 (Figure8a,b,c) and 12:02:00 to12:04:08 UT (Figure 8c,d,e) on 11 April 2010, with a frequency resolution of 0.0078 Hz. The time intervals in Figure 8a,b correspond to those shown in Figure 7a,b, respectively. In Figure 8a,b and Figure 8d,e, we can see a continuous high-PSD band at frequencies of 0.4 to 0.6 Hz in both time intervals. The coherence between the H component at ATH and the D component at MGD is close to one in the frequency range of 0.4 to 0.6 Hz.

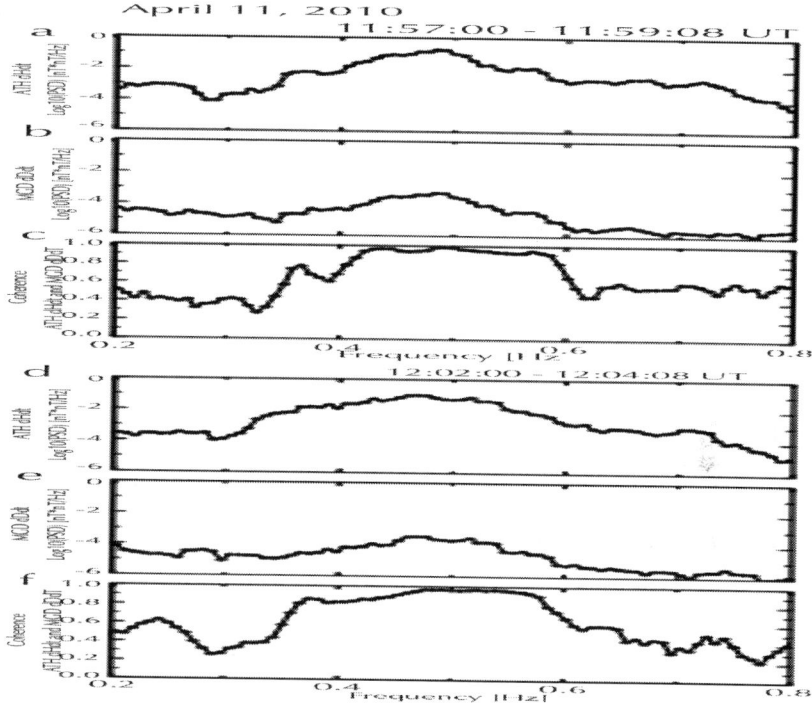

Figure 8: The PSD and the coherence of Pc1 waveforms of the H and D components in case 2. Upper three panels show the power spectrum density of (a) H component of the magnetic field at ATH and (b) D component of the magnetic field at MGD, as well as (c) coherence of Pc1 waveforms between H component at ATH and D component at MGD, observed at 11:57:00 to 11:59:08 UT on 11 April 2010, for a frequency range of 0.2 to 0.8 Hz. Lower three panels (d-f) show the same quantities observed at 12:02:00 to 12:04:08 UT on 11 April 2010.

Figure 9a,b shows the cross-correlation between the upper envelopes of ATH H and MGD D components of Pc1 pearl structure at 11:57:00 to 11:59:08 and 12:02:00 to 12:04:08 UT, respectively, on 11 April 2010, in the same format as Figure 5. In Figure 9a,b, we can see that the correlation is greater than 0.8 with a time difference of approximately 4 s during both time intervals. As shown in Figure 6m, we made this cross-correlation analysis of Pc1 envelopes for the whole time interval. The cross-correlation of Pc1 envelopes between the H component at ATH and the D component at MGD is greater than 0.8 throughout this Pc1 event, which also has high coherence ($r > 0.8$).

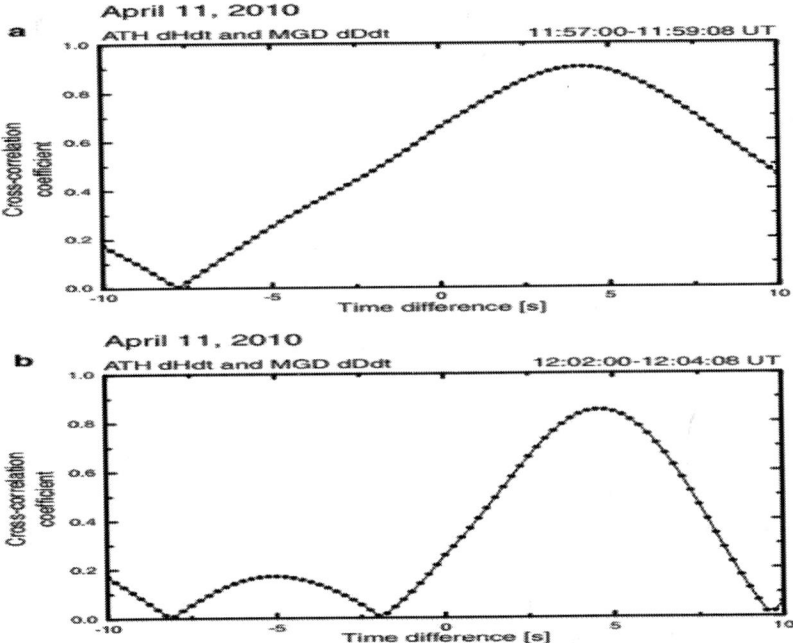

Figure 9: The cross-correlation analysis in case 2. Cross-correlation between H component at ATH and D component at MGD obtained for upper envelope of Pc1 pearl structures at (a) 11:57:00 to 11:59:08 and(b) 12:02:00 to 12:04:08 UT on 11 April 2010.

DISCUSSION

Comparing two case studies, we found that the Pc1 pearl structures observed at widely separated ground stations can be generally similar. However, case 1 shows that detailed pearl structures are different in some time intervals, even if the coherence of Pc1 waveforms between two different stations is close to one. The polarization angle varied depending on frequency for case 1, suggesting a spatially distributed ionospheric source. On the other hand, case 2 shows that the coherence of Pc1 waveforms and cross-correlation of Pc1 envelopes can both be high ($r > 0.8$). In this second case, the polarization angle was almost constant for frequencies from 0.4 to 0.6 Hz. Here, we discuss the mechanisms that may have contributed to these differences.

Possible Generation Mechanisms of Pc1 Pearl Structures in the Magnetosphere

One of the possible generation mechanisms is based on the BWP model (e.g., Guglielmi et al. ([1996]); Mursula et al. ([1999])). This model explains that Pc1 pearl structures are caused by bouncing of Pc1 waves along the geomagnetic field line between the northern and southern hemispheres. According to the BWP model, the repetition period of Pc1 pearl structures would be related to the length of the magnetic field line, as well as to the Alfven velocity in the magnetosphere. According to this model, the expected repetition period of Pc1 pearl structures is several tens of seconds, depending on the radial distance of the generation region of EMIC waves located near the magnetic equator. However, some studies have reconsidered the BWP model, because they found that the observations did not match the expected results from this model. For example, since the BWP model is based on comparison of ground and satellite data, Perraut ([1982]) found that the repetition period of Pc1 pearl structures seen on the ground station, did not clearly match the one observed in space. In addition, Erlandson et al. ([1990]) measured the Poynting flux of EMIC waves using the Viking satellite to investigate Pc1 pulsations near the plasmapause. They found that the energy flux of Pc1 pearl structures was mainly downward, along the magnetic field line. Moreover, Paulson et al. ([2014]) measured the average wave power over 0.6 to 0.8 Hz of Pc1 waves observed at the Hornsund station on the ground and the Van Allen probe in space. They found that both repetition periods of an average wave power were approximately 130 s. They suggested that the similar repetition periods on the ground and in space contradict the BWP model, because if the BWP model is correct, the repetition period of average wave power in space would have to be half of that observed on the ground. In Figures 3and 7, the repetition period of Pc1 pearl structures at three stations was approximately 10 s, which is shorter than the expected repetition period from the BWP model. Mursula et al. ([2001]) and Mursula ([2007]) attempted to explain the Pc1 pearl structures as the result of modulation of EMIC waves by long-period ULF waves (such as Pc4 to 5 pulsations). The repetition periods of Pc1 pearl structures found in this study, approximately 10 s in case 1, and approximately 10 to 40 s in case 2, are shorter than the period of Pc4 to 5 pulsations. Such generation mechanisms of

Pc1 pearl structures in the magnetosphere are not able to explain the different wave structures at different stations that we observed, even in the case of Pc1 pulsations that propagated from the same source. If Pc1 waves with different frequencies are mixed in the magnetosphere, these waves should have similar waveforms, even if they are detected at different stations. As we observed in case 1, the detailed Pc1 pearl structures were slightly different at the three stations. In Figure 2m, the cross-correlation of Pc1 envelopes at ATH and MGD is less than 0.5 in the time interval of the first and third Pc1 bursts, although the coherence of Pc1 waveforms is close to one.

Comparison of Observations and Model Calculations of Pc1 Pearl Structures

Some studies have considered that Pc1 pearl structures can be caused by beating in the ionosphere. This is the consequence of amplitude modulation of Pc1 waves caused by superposition of the waves at slightly different frequencies during their propagation through the ionospheric duct (Pope [1964]). Nomura et al. ([2011]) found that some Pc1 events observed at low latitudes have a polarization angle that is frequency dependent. This indicates that these Pc1 pulsations have a spatially distributed source region in the ionosphere that can cause beating in the ionosphere to create Pc1 pearl structures. In case 1, Figure 2g,h,i shows that Pc1 pulsations at three different stations have a polarization angle that is dependent on frequency, and thus we can suggest that these waves have a spatially distributed source in the ionosphere. Figure 3a,b shows that the Pc1 pearl structures varied with a repetition period of 10 s, suggesting that these structures may also be caused by beating in the ionosphere. Moreover, their amplitude envelopes are slightly different at the three stations. On the other hand, in case 2, the polarization angle does not show any dependence on frequency (Figure 6g,h,i), indicating that these waves have a localized ionospheric source. Figure 7a,b shows that the Pc1 pearl structures at the three stations are similar, with a repetition period of approximately 10 s.

To understand the different features of Pc1 pearl structures between the two case studies, we did two model calculations of Pc1 pearl structures under two different conditions, as shown in Figure10. We generated simple sinusoidal waves in the frequency range of 0.6 to 1.0 Hz, and assumed that they propagated through the ionospheric duct with an Alfven velocity of 500 km/s, as estimated by Fraser ([1975]). We took into account the time difference and polarization angle variation due to the relative location of these wave sources and the observation points. We did not consider the mode conversion effect and polarization sense of Pc1 waves during their propagation through the ionospheric duct. During the duct propagation, we consider an attenuation effect that causes the amplitude of Pc1 waves to decrease by 25 % per 1,000 km distance from the source region to each station. Subsequently, we checked the pearl structures under two different conditions. First, model 1 assumes that the Pc1 waves are generated at a north-south extended source region (orange line in Figure 10a) with frequencies from 0.6 (high latitude) to 1.0 Hz (low latitude) at different latitudes. This would correspond to a pearl structure caused by beating during duct propagation in the ionosphere, from their source points to the observation points. Second, model 2 takes into account that Pc1 waves with different frequencies are mixed at a point source in the ionosphere (orange point in Figure 10b), corresponding to pearl structures created in the magnetosphere.

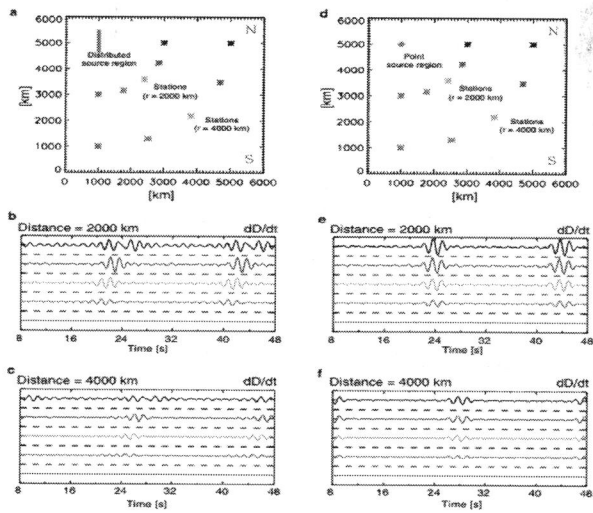

Figure 10: Simple model calculations for comparison between a distributed source and a point source. Location of stations and source region: (a) A dis-

tributed source region (model 1) and (d) a point source region (model 2). The D component waveforms of the source waves with frequencies of 0.6 to 1.0 Hz: distance from the source region to stations at (b) 2,000 and (c) 4,000 km for model 1 and (e) 2,000 and (f) 4,000 km for model 2. Colors indicate angles of stations from the south.

Figure 10a,b,c and Figure 10d,e,f show the results of models 1 and 2, respectively. For model 1, the source region is distributed from north to south with a length of 1,000 km. The waveforms of Pc1 waves in Figure 10b,c show that Pc1 pearl structures are slightly different at different stations, particularly for a station located at 90° (black dot in Figure 10a and black lines in Figure 10b,c), corresponding to a perpendicular direction from the source distribution. Additionally, the time difference between two stations with the same distance from the source region varies because of the changing angle of the stations from the south. In the case of model 2, as shown in Figure10d,e,f, Pc1 waves coming from a point ionospheric source have identical waveforms at different stations. The time differences between two different stations with the same distance from a source region are close to zero. If the Pc1 waves propagated from a spatially distributed source region in the ionosphere, the different Pc1 pearl structures would be observed at different stations due to beating processes in the ionosphere, even though the waves are coming from the same ionospheric source region. In case 1, we found that Pc1 pearl structures were slightly different in some time intervals, with high coherence of Pc1 waveforms. We also found that the variation of polarization angle at the three stations depended on frequency. In case 2, however, the Pc1 pearl structures were similar, and the polarization angle was independent of frequency. As shown in the model calculations of Pc1 waves, we suggest that the observed case 1 could be caused by beating processes in the ionosphere, while Pc1 pearl structures in case 2 could be created by magnetospheric effects.

In addition, we cannot exclude the possible effects that dispersive propagation could have on ducted Pc1 waves. The effect can also contribute to the formation of Pc1 pearl structures in the ionosphere. Because the group velocity of dispersive waves differs from the phase speed, it can cause the modulation of wave amplitude in a wave packet. The high-latitude transmission and reflection properties of the ionosphere in the Pc1 frequency range is related to the wave number and the wave vector   (Greifinger [1972]). As shown by model calculations by Fujita ([1987],[1988], the group velocity of Pc1 pulsations as a function of frequency increases near the lower cut off frequency. If the observed Pc1 waves have a broad bandwidth, the amplitude modulation of Pc1 waves could be caused by dispersive propagation through the ionospheric duct. From our observations, Pc1 pearl structures can have different shapes at different stations (case 1) and similar shapes at different stations (case 2). The bandwidth of case 1 (approximately 0.5 Hz) was wider than that of case 2 approximately, suggesting that dispersive propagation contributes more to the creation of Pc1 pearl structures in the first case. However, in this study, we cannot quantify the contribution of this effect to the creation of Pc1 pearl structures in the ionosphere.

CONCLUSIONS

From 1 January 2009 to 31 December 2011, we investigated pearl structures of Pc1 geomagnetic pulsations observed by induction magnetometers at three mid- to low-latitude ground stations (ATH, MGD, and MOS). We selected two Pc1 pulsation events observed simultaneously at three stations: case 1 on 8 April 2010 and case 2 on 11 April 2010. The results of this study can be summarized as follows:

- For case 1, even though the coherence of Pc1 waveforms at different stations is high, the Pc1 pearl structures were slightly different at different stations in some time intervals. The polarization angle varied depending on frequency, indicating that the Pc1 pulsations propagated from a spatially distributed ionospheric source.

- For case 2, the Pc1 pearl structures are similar at different stations with high coherence of Pc1 waveforms. The polarization angle was almost constant, indicating that the source region of Pc1 pulsation is positioned in a localized region in the high-latitude ionosphere.

- Pc1 pearl structures with a repetition period of approximately 10 s in case 1 and approximately 10 to approximately 40 s in case 2 were observed at three stations. These periods are shorter than those expected based on the BWP model.

- From the model calculation of Pc1 pearl structures, we found that the pearl structures propagating from an ionospheric point source should have identical waveforms at different stations. The pearl structures generated by beating in the ionosphere with a spatially distributed source can be different at different stations.

From these results, we suggest that beating processes in the ionosphere with a spatially distributed ionospheric source can cause pearl structures during the ionospheric duct propagation from high to low latitudes, with long distances from the source to the stations. In case 2, however, we cannot reliably interpret the Pc1 pearl structures with a constant polarization angle using the beating process in the ionosphere. Therefore, we cannot exclude the possibility that mechanisms in the magnetosphere also contribute to the generation of Pc1 pearl structures. In order to understand and quantify the contribution of beating in the ionosphere to the creation of Pc1 pearl structures, we would like to further investigate the statistical characteristics in future studies.

AUTHORS' CONTRIBUTIONS

CWJ carried out the spectral analysis for case 1 and case 2, prepared simple model calculations for comparison between a distributed source and a point source, and wrote the manuscript. KS helped in planning for the design of the study and interpretation and drafted the manuscript as the supervisor of CWJ. MC and IS carried out the induction magnetometer observation at ATH in Canada and drafted the manuscript. IP and BS carried out the induction magnetometer observation at MGD in Russia and drafted the manuscript. All authors read and approved the final manuscript.

ACKNOWLEDGEMENTS

We thank M. Sera and Y. Ikegami at the Moshiri observatory of the Solar-Terrestrial Environment Laboratory, Nagoya University, all the staff of the Institute of Cosmophysical Research and Radiowave Propagation (IKIR), and Y. Katoh, H. Hamaguchi, and Y. Yamamoto of STEL, Nagoya University, for their help and support in the operation of the induction magnetometers. The Dst and AE indices were provided by the WDC-C2 for geomagnetism at Kyoto University. This work was supported by Grants-in-Aid for Scientific Research (16403007, 18403011, 19403010, and 20244080), the 21th Century COE Program (Dynamics of the Sun-Earth-Life Interactive System, No. G-4), the Global COE Program of Nagoya University 'Quest for Fundamental Principles in the Universe (QFPU)', the Special Funds for Education and Research (Energy Transport Processes in Geospace), and the IUGONET Project from MEXT, Japan, as well as the Leadership Development Program for Space Exploration and Research from Nagoya University for Leading Graduate Schools.

REFERENCES

1.	Anderson BJ, Erlandson RE, Zanetti LJ (1992) A statistical study of Pc 1-2 magnetic pulsations in the equatorial magnetosphere: 1. equatorial occurrence distributions. J Geophys Res Space Phys 97:3075-3088

2.	Campbell WH (1967) Low attenuation of hydromagnetic waves in the ionosphere and implied characteristics in the magnetosphere for Pc 1 events. J Geophys Res 72(13):3429-3445

3.	Erlandson R, Zanetti L, Potemra T, Block L, Holmgren G (1990) Viking magnetic and electric field observations of Pc 1 waves at high latitudes. J Geophys Res Space Phys 95(A5):5941-5955

4.	Fowler R, Kotick B, Elliott R (1967) Polarization analysis of natural and artificially induced geomagnetic micropulsations. J Geophys Res 72(11):2871-2883

5.	Fraser B (1975) Ionospheric duct propagation and Pc 1 pulsation sources. J Geophys Res 80(19):2790-2796

6.	Fujita S (1987) Duct propagation of a short-period hydromagnetic wave based on the international reference ionosphere model. Planet Space Sci 35(1):91-103

7.	(1988) Duct propagation of hydromagnetic waves in the upper ionosphere, 2, dispersion characteristics and loss mechanism. J Geophys Res Space Phys 93(A12):14674-14682

8.	Fukunishi H, Toya T, Koike K, Kuwashima M, Kawamura M (1981) Classification of hydromagnetic emissions based on frequency-time spectra. J Geophys Res Space Phys 86(A11):9029-9039

9.	Greifinger P (1972) Ionospheric propagation of oblique hydromagnetic plane waves at micropulsation frequencies. J Geophys Res 77(13):2377-2391

10. Guglielmi A, Feygin F, Mursula K, Kangas J, Pikkarainen T, Kalisher A (1996) Fluctuations of the repetition period of Pc1 pearl pulsations. Geophys Res Lett 23(9):1041-1044

11. Jacobs J, Watanabe T (1964) Micropulsation whistlers. J Atmos Terr Phys 26(8):825-826

12. Kawamura M, Kuwashima M, Toya T (1981) Comparative study of magnetic Pc1 pulsations between low latitudes and high latitudes: source region and propagation mechanism of the waves deduced from the characteristics of the pulsations at middle and low latitudes. Mem Natl Inst Polar Res. Special issue 18:83-100

13. Kim H, Lessard M, Engebretson M, Young M (2011) Statistical study of Pc1-2 wave propagation characteristics in the high-latitude ionospheric waveguide. J Geophys Res Space Phys 116(A7):07227

14. Kuwashima M, Toya T, Kawamura M, Hirasawa T, Fukunishi H, Ayukawa M (1981) Comparative study of magnetic Pc1 pulsations between low latitudes and high latitudes: statistical study. Mem Natl Inst Polar Res. Special issue 18:101-117

15. Manchester R (1966) Propagation of Pc 1 micropulsations from high to low latitudes. J Geophys Res 71(15):3749-3754

16. Mursula K (2007) Satellite observations of Pc 1 pearl waves: the changing paradigm. J Atmos Sol Terr Phys 69(14):1623-1634

17. Mursula K, Rasinkangas R, Bösinger T, Erlandson R, Lindqvist P-A (1997) Nonbouncing Pc 1 wave bursts. J Geophys Res Space Phys 102(A8):17611-17624

18. Mursula K, Kangas J, Kerttula R, Pikkarainen T, Guglielmi A, Pokhotelov O, Potapov A (1999) New constraints on theories of Pc1 pearl formation. J Geophys Res Space Phys 104(A6):12399-12406

19. Mursula K, Bräysy T, Niskala K, Russell C (2001) Pc1 pearls revisited: structured electromagnetic ion cyclotron waves on polar satellite and on ground. J Geophys Res Space Phys 106(A12):29543-29553

20. Nomura R, Shiokawa K, Pilipenko V, Shevtsov B (2011) Frequency-dependent polarization characteristics of Pc1 geomagnetic pulsations observed by multipoint ground stations at low latitudes. J Geophys Res Space Phys 116(A1):01204

21. Nomura R, Shiokawa K, Sakaguchi K, Otsuka Y, Connors M (2012) Polarization of pc1/EMIC waves and P related proton auroras observed at subauroral latitudes. J Geophys Res Space Phys 117(A2):02318

22. Obayashi T (1965) Hydromagnetic whistlers. J Geophys Res 70:1069-1078

23. Paulson K, Smith C, Lessard M, Engebretson M, Torbert R, Kletzing C (2014) In situ observations of Pc1 pearl pulsations by the van allen probes. Geophys Res Lett 41(6):1823-1829

24. Perraut S (1982) Wave-particle interactions in the ulf range: geos-1 and-2 results. Planet Space Sci 30(12):1219-1227

25. Pope JH (1964) An explanation for the apparent polarization of some geomagnetic micropulsations (pearls). J Geophys Res 69(3):399-405

26. Rasinkangas R, Mursula K (1998) Modulation of magnetospheric emic waves by Pc 3 pulsations of upstream origin. Geophys Res Lett 25(6):869-872

27. Sakaguchi K, Shiokawa K, Miyoshi Y, Otsuka Y, Ogawa T, Asamura K, Connors M (2008) Simultaneous appearance of isolated auroral arcs and Pc 1 geomagnetic pulsations at subauroral latitudes. J Geophys Res Space Phys 113(A5):05201

28. Shiokawa K, Nomura R, Sakaguchi K, Otsuka Y, Hamaguchi Y, Satoh M, Katoh Y, Yamamoto Y, Shevtsov B, Smirnov S, Poddelsky I, Connors M (2010) The STEL induction magnetometer network for observation of high-frequency geomagnetic pulsations. Earth Planets Space 62(6):517

29. Tepley L, Landshoff R (1966) Waveguide theory for ionospheric propagation of hydromagnetic emissions. J Geophys Res 71(5):1499-1504

30. Troitskaya V, Gul'Elmi A (1967) Geomagnetic micropulsations and diagnostics of the magnetosphere. Space Sci Rev 7(5–6):689-768

31. Usanova M, Mann I, Rae I, Kale Z, Angelopoulos V, Bonnell J, Glassmeier K-H, Auster H, Singer H (2008) Multipoint observations of magnetospheric compression-related EMIC Pc1 waves by themis and carisma. Geophys Res Lett 35(17):17-25

32. Waters C, Lysak R, Sciffer M (2013) On the coupling of fast and shear Alfvén wave modes by the ionospheric hall conductance. Earth Planets Space 65(5):385-396

Citations

CHAPTER 1

F. Dudkin, Gautam Rawat, B. R. Arora, V. Korepanov, O. Leontyeva, and A. K. Sharma, Application of Polarization Ellipse Technique for Analysis Of ULF Magnetic Fields from Two Distant Stations in Koyna-Warna Seismoactive Region, West India, doi:10.5194/nhess-10-1513-2010.

CHAPTER 2

X. Zhang, X. Shen, M. Parrot, Z. Zeren, X. Ouyang, J. Liu, J. Qian, S. Zhao, and Y. Miao, Phenomena of Electrostatic Perturbations before Strong Earthquakes (2005–2010) Observed on DEMETER, doi: 10.5194/nhess-12-75-2012.

CHAPTER 3

Masashi Hayakawa, Possible Electromagnetic Effects on Abnormal Animal Behavior before an Earthquake, doi: 10.3390/ani3010019.

CHAPTER 4

Vittorio Sgrigna and Livio Conti, "A Deterministic Approach to Earthquake Prediction," International Journal of Geophysics, vol. 2012, Article ID 406278, 20 pages, 2012. doi:10.1155/2012/406278.

CHAPTER 5

Gabriel Abadal, Javier Alda and Jordi Agustí (2014) Electromagnetic Radiation Energy Harvesting – The Rectenna Based Approach, ICT - Energy - Concepts Towards Zero - Power Information and Communication Technology, Dr. Giorgos Fagas (Ed.), ISBN: 978-953-51-1218-1, InTech, DOI: 10.5772/57118.

CHAPTER 6

K. Ivanovich, S. Evgenyevich, G. Vasilyevich, D. Nikolaevna and V. Igorevich, "Features of Usage of Electromagnetic Field of Extremely Low Frequency for the Storage of Agricultural Products," *Journal of Electromagnetic Analysis and Applications*, Vol. 5 No. 5, 2013, pp. 236-241, doi: 10.4236/jemaa.2013.55038.

CHAPTER 7

A. Memmedov, T. Abbasov and M. Şeker, «Theoretical Modeling and Experimental Analysis of Drying Process in Electromagnetic Field,» World Journal of Engineering and Technology, Vol. 2 No. 1, 2014, pp. 41-53, doi:10.4236/wjet.2014.21005.

CHAPTER 8

Ryuho Kataoka, Yoko Fukuda, Yoshizumi Miyoshi, Hiroko Miyahara, Satoru Itoya,Yusuke Ebihara, Donald Hampton, Hanna Dahlgren, Daniel Whiter and Nickolay Ivchenko, Compound Auroral Micromorphology: Ground-based High-speed Imaging, doi:10.1186/s40623-015-0190-6.

CHAPTER 9

Chae-Woo Jun, Kazuo Shiokawa, Martin Connors, Ian Schofield, Igor Poddelsky, and Boris Shevtsov, Study of Pc1 Pearl Structures Observed at Multi-Point Ground Stations in Russia, Japan, and Canada, doi:10.1186/s40623-014-0140-8.

Index

A

Analog-digital converter (ADC) 167
Analog sensors (AS) 167
Application-Specific Integrated Circuit (ASIC) 139

B

Biological effect 56
Biological systems 160

C

Chemical propertie 176, 181, 183
Command unit (CU) 102
Continuous-Wave (CW) 137

D

Data acquisition and processing system (DAPS) 102

Digital-analog converter (DAC) 167

E

Earthquake (EQ) 43, 46
Earthquakes (EQs) 2, 44
Electromagnetic (EM) 34, 120
Electromagnetic ion cyclotron (EMIC) 201, 218
Electromagnetic wave 120, 152
Electromagnetic wave (EMW) 176, 177
Equatorial Ionospheric Anomaly (EIA) 36
European space agency (ESA) 97
Extremely-low-frequency (ELF) 52

F

Fast Fourier transform (FFT) 206, 222

Field-aligned electron bursts (FABs)
    210
Field-of-view (FOV) 203
Food product 165
Food production 174

G

Gradient magnetic field (GMF) 158
Grapheme, the field effect transistor
    (G-FET) 150

H

High frequency and low power
    (HFLP) 146

I

Inertial Alfvén waves (IAWs) 201
Information and Communications
    Technology (ICT) 120
International space station (ISS) 97

L

Lithosphere-atmosphere-ionosphere
    (LAI) 24
Low-earth-orbit (LEO) 64, 69

M

Magnetometer head (MH) 98
Massachussetts Institute of Technol-
    ogy (MIT) 137
Matching amplifier (MA) 167
Metal-Insulator-Insulator-Metal
    (MIIM) 131, 147
Metal-insulator-metal (MIM) 130
Microorganism 158, 174

N

National Renewable Energy Labora-
    tory (NREL) 127

O

Optical antenna (OA) 130

Optical rectenna (OR) 129

P

Physical process 178
Poker Flat Research Range (PFRR)
    200, 203
Polarization ellipse (PE) 4
Polarization ellipses (PE) 2
Power spectrum density (PSD) 222
Power supply system (PSS) 102
Precision audio oscillator of low
    frequency (PAOLF) 166

R

Radio Frequency IDentification
    (RFID) 137
Radiofrequency rectenna (RFR) 129
Random access memory (RAM) 167
Raytheon Airborne Microwave Plat-
    form (RAMP) 137

S

Scientific complementary metal-
    oxide semiconductor (sCMOS)
    200, 203
Seismo electro -magnetic emissions
    (SEME) 69
Single-chip microcomputer (SCMC)
    167
Solar-power satellite (SPS) 137
South atlantic anomaly (SAA) 85

T

Transverse electric (TE) 34

U

Ultra low frequency (ULF) 1, 2
Ultra-low-frequency (ULF) 50

W

Water sample 158
Wide range 164, 165